INTRODUCING
CELLULAR COMMUNICATIONS
THE NEW MOBILE TELEPHONE SYSTEM

Other Recent Books by the Author:

No. 1532 *The Complete Book of Oscilloscopes*
No. 1632 *Satellite Communications*

INTRODUCING CELLULAR COMMUNICATIONS THE NEW MOBILE TELEPHONE SYSTEM

BY STAN PRENTISS

TAB BOOKS Inc.
BLUE RIDGE SUMMIT, PA. 17214

FIRST EDITION

FIRST PRINTING

Copyright © 1984 by TAB BOOKS Inc.

Printed in the United States of America

Reproduction or publication of the content in any manner, without express permission of the publisher, is prohibited. No liability is assumed with respect to the use of the information herein.

Library of Congress Cataloging in Publication Data

Prentiss, Stan.
   Introducing cellular communications.

   Includes index.
   1. Cellular radio. I. Title.
TK6570.M6P74   1984   384.5'3   83-24186
ISBN 0-8306-0682-3
ISBN 0-8306-1682-9 (pbk.)

Cover photography courtesy of Ameritech Mobile Communications.

Contents

Introduction   vi

1 The What and Why of Cellular Communications   1
The Birth of an Industry—System Concepts—An AMPS Cell

2 Terrestrial and Satellite Carriers   27
AMPS in Chicago—Western Union in Buffalo—NASA

3 Pertinent Specifications   61
Limitations on Emissions—Land/Base Station Transmitter—Land/Base Station Receiver—Mobile Transmitters and Receivers—Mobile Call Processing—Signaling Formats—Definitions

4 Available Systems and Equipment   99
Motorola's DYNA T•A•C—OKI's Control and Transceiver Units — Anaconda-Ericsson System — General Electric and Northern Telecom System—ITT's Celltrex—E.F. Johnson Equipment—Fujitsu Ten's AVM

5 General Electric's CB-Telephone Proposal   153
The User Market—PSTN Interconnect—System Operation—Proposed Parameters—Signaling Formats

6 Test Programs and Equipment   185
Test Equipment Review—Motorola Diagnostics—Electronic Accessories

Index   213

Introduction

Effectively, this book is both a primer and an advanced technical information/discussion of a somewhat complex and very new subject that, when fully operational, will make its investors virtual partners in Fort Knox.

The idea of cruising across the land conducting private, public, and business conversations with anyone virtually anywhere in the world is staggering to most imaginations. Fortunately, the concept is very real, and a number of cellular radio systems are already in operation in Chicago, Washington-Baltimore, and possibly North Carolina. Later there will be two systems—one telephone-type company wireline, and another nontelephone type called nonwireline to serve you wherever you travel in the major 90 cities throughout the U.S. In the rural areas, NASA would like to take over the job of cellular distribution so the few out there might be served by satellite with equal access to all other systems.

Whatever the ultimate outcome, this is no more than another branch of the communications explosion that could unite us all, eventually, into one world, or at least spread English and Spanish around a little more than at present. Perhaps buns instead of bombs could become international topics when we can speak freely with them. At any rate, the spoken word now goes completely mobile, and such terms as MTSOs and cell sites, handsoff, and SMSAs gleefully enter your consciousness.

If you're consumed with curiosity—as millions will soon

be—do read the book and discover what your next dinner conversation piece is all about. Cellular Radio has indeed arrived!

The most recent information is provided on initial FCC system approvals, all pertinent system and user statistics, existing radio setups, plans for satellite carriers, and even a pseudo-cellular system now under development by General Electric for 900 MHz FM service at cost figures approaching those of CB (Citizens Band Radio) during the 1970s.

Included are all Federal specifications, signaling requirements, discussions of such novel basics such as cell sites, handoffs, SATs, Carey Curves, traffic analyses, sector antennas, MSAT, reverse voice channels, and mobile station/land station compatibility. And even Sweden gets into the act with its Anaconda-Ericsson system that's been operational in Europe at 450 MHz for the past several years.

You'll learn that the U.S. system has 666 channels with one wireline and one non-wireline cellular unit permitted for each of the 90 major cities and that the National Aeronautics and Space Administration is trying to put all systems in the boonies on satellite to join the remainder, wherever they may be.

Then you'll discover that in the beginning, $2,500 to $3,000 buys one of these expensive mobiles, and routine costs could add up to as much as $170 per month additional if you're talkative. But within several years such prices will moderate, and the expectations are that by 1990, such mobiles and handhelds will cost less than $1,000—at least that's the going prediction. Nonetheless, cellular radio does open international telephone service to anyone at home or on the road during day or night, wherever there's a cell site or another mobile to talk to.

I would also like to acknowledge and express my grateful appreciation to the following for their aid and forebearance in helping this publication to attain both content and covers:

Kevin Colosia, Normal Bach, David Wiesz, Knobby Clark, Motorola Communications; Bill Adler, Steve Markendorf, Mike Ferrante, Wendell Harris, Jay Kawalski, John Burkowski, John Reed, Janis Langley, and Peggy White, Federal Communications Commission; John Kese, James Kearney, General Electric (Syracuse); Guy Pierce, Western Union; Linda Urben, Margaret Cathcart, AMPS (Basking Ridge); Jerry Freibaum, NASA (Headquarters); Joseph Kobylak, OKI; Douglas Cook, Richard Lee, Cushman; Ed Tingley, Peter Bennett, Electronic Industries Association; Gunnar Svala, Anaconda Erickson.

Chapter 1

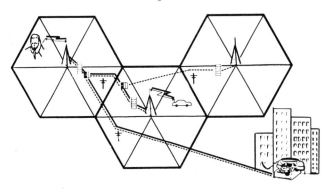

The What and Why of Cellular Communications

Mobile radio service actually began in 1921 when Detroit police operated an experimental system at 2 MHz. Today, over 40 MHz of spectrum between 30 MHz and 500 MHz has been allocated by the FCC for all types of users. Now, for cellular communications alone, the FCC is allocating 40 MHz of spectrum between 825-845 MHz and 870-890 MHz for wireline and non-wireline vendors. The size of this allocation indicates how important this new means of communications will become. And, in addition to 40 channels of Citizens Band Radio around 27 MHz, another application for psuedo-cellular CB around 900 MHz is pending, having already been released by the FCC for reply and comment.

Until the mid-1960's mobile radio service was strictly push-to-talk—that is, manually operated. But in 1964, an automatic channel selection named MJ was put into operation at 150 MHz. Thereafter, subscribers had only to dial whatever specific numbers they wished. Then in 1969, an MK system in the 450 MHz spectrum was installed in the Improved Mobile Telephone System. This system set the present standard for existing mobile service.

Today, mobile telephone service requires land transmitter stations located at either high elevations or with tall towers that generate outputs of some 250 watts and antenna gains that double the effective radiating power (ERP) to 500 watts. Coverage supplied by such stations extends between 20 and 25 miles, but frequency interference with stations considerably beyond this range often occurs. This obviously limits the number of available stations as well as the frequencies on which they are permitted to operate. Consequently, such mobiles and their base stations are relative luxuries, costing as much as 15 to 20 times over standard telephone rates. That, naturally, makes mobile service something of a luxury, and, with limited spectrum, permits only a few users in categories where expeditious communications justify considerable cost.

In 1949, Radio Common Carriers (RCCs) were established to supply equivalent service in competition with wireline facilities such as the Bell System. As of 1977, there were 1,351 Bell and independent wireline carriers and 1,375 radio common carriers. At the end of 1982, the FCC lists 2,481 wireline stations (call signs) and 2,574 non-wireline stations. These figures have almost doubled in a short 5-year period, and they will continue to expand as service demand grows. However, there is a limit to the available spectrum and the number of subscribers it can support. Therefore, cellular communications was born.

THE BIRTH OF AN INDUSTRY

In 1974, the FCC eliminated UHF TV channels 70 through 83 and other sectors in the UHF spectrum between 806 MHz and 947 MHz. The first 40 MHz was allocated to wireline common carrier and another 30 MHz was allocated for additional private services. The remainder was reserved for future needs. But in 1975, the FCC modified its earlier decision and opened the 40 MHz allocation to "any qualified common carrier."

With a total of 115 MHz now available for mobile service, cellular planning—begun as long ago as 1947—commenced in earnest but with several prime differences over conventional mobile radio.

The New Approach

There was still to be both mobile and handheld units as well as land stations as the prime receivers, but general service areas could be divided into spectrum reusable cells that would use less power but actually cover a wider distribution area. This was to be done via an automatic system of *handoffs* and multichannel selections as well as automatic signal sampling. For when mobiles on the move change locations rapidly it would be highly advantageous to be able to travel from one cell to another without having a conversation interrupted. Handoffs do exactly that. Therefore, in large cells, omnidirectional antennas are used to cover a larger territory. In closely spaced cells, directional antennas are used to divide cells into as many as three divisions, reducing cochannel interference and increasing overall system efficiency. This approach encourages flexibility in expanding systems yet remains sufficient to meet small system needs at startup.

Another factor entering the picture is one of nationwide compatibility. Since FCC and EIA specifications require that transceivers all perform within certain specifications and all have equivalent outputs, there are only certain ways to accomplish such results. Consequently, one mobile should pretty well sound like another and should be usable in any similar cellular setup in the U.S. without the slightest modification. So the standardization which has been sadly lacking in so many communications systems in the past is now a Federal requirement that has to be met as both a legal and electronic requirement for FCC type acceptance and commercial application. Therefore, those manufacturers who can build acceptable cellular transceivers have an equal chance in the marketplace, regardless of their national or international antecedents.

The FCC Ruling

On April 9, 1981, the Federal Communications Commission approved the concept of cellular mobile telephone radio service, and on December 8, 1982, it announced a construction permit for Advanced Mobile Phone Service, Inc. (AMPS—of which you will hear a great deal more) for the city of Buffalo, New York. AMPS is a wholly-owned subsidiary of the American Telephone and Telegraph Co. (AT&T). Approval of this request follows November approval of similar AMPS wireline systems for Chicago, Pittsburgh, and Boston. Historically, of course, this sets the precendent, and others in the top 30, 60, and 90 cities are expected to follow rapidly, with non-wireline (other than phone companies) taking the longest since they are all subject to comparative hearings before the FCC makes further assignment determinations.

Thus far, 52 wireline and 142 non-wireline applicants (a total of 194) have filed for the top 30 cities, followed by some 396 for the second tier of 30 cities (31 through 60). The third group (61 through 90) due on March 8, 1983, amounted to a total of 567 since these are the smaller cities and less elaborate cell sites, and therefore less expensive installations, are required. Finally, by December 1, 1983, anyone else throughout the U.S. may file with the FCC, addressing such an application to the Office of the Secretary, Room

222, Federal Communications Commission, 1919 M St., N.W., Washington, DC 20554. All must abide by orders such as 90 FCC 82-466, 47 Fed. Reg. 32537 and FCC 82-566, or 89 FCC 2nd 58, 47 Fed. Reg. 10018 (1982) and Mimeo 2973, with any revisions.

With the exception of New England, which has a County Metropolitan Areas (NECMA) limit, Cellular Geographic Service Areas (CGSA) may not extend beyond the boundaries of Standard Metropolitan Statistical Areas (SMSAs) unless the extension is either de minimis or does not include "area within another central SMSA." Terms such as central and de minimis are explained in the FCC's public notice of March 24, 1982, if you wish to dig deeper. Initial cellular application filing procedures are contained in FCC Mimeo 2973, dated 24 March 1982, and 89 FCC 2d 58, 47 Fed. Reg 10018 (1982). There is also a Nov. 1, 1982 update by the Common Carrier Bureau available to those interested. But by now, of course, all major filings have taken place and the majority of applications have been either approved or rejected. So I am mainly quoting history for the record. CGSA applications, nonetheless, are to be drawn on one or more U.S. Geographical Survey maps on a scale of 1:250,000. Scales, latitudes and longitudes, CGSA boundaries, base station sites, and the 39 dBu contours of each cell (including their numbers and/or other identification) are to be clearly marked. Microwave links connecting cell sites do not have to be included for basic qualifications.

Corporate Interest

Telephone land lines versus microwave and fiber optics is even now the big divider and/or equalizer between AMPS and other non-telephone companies, which can usually be called independents. Radio/tv broadcasters and other groups already in the communications business are already joining cellular pioneers in collective ventures that could pay off handsomely as the service and its equipment matures. Big broadcast towers, of course, are naturals for microwave dishes and message transceiving of all descriptions with very little additional costs. Further, the prospect of large data transmissions at considerable savings over conventional methods in off-hours could make news and financial institutions close partners in most, if not all, cellular undertakings. So, already, considerable aims and ends are becoming apparent to separate the dual community services wisely specified by the FCC for all urban locations.

Partnerships, however, are not restricted to the independents. Consortiums of AT&T, GTE, and even Western Union (WU) are already evident among the wireline group. They can be expected to offer their own brand of spinoff services which may or may not complement those of the independents. At this early stage in the overall development, unfortunately, there are few facts available other than educated guesses. And these are not always practical. Certainly computer and allied print material transmissions may be expected in addition to voice and other communications media. The cellular people, of course, see most applications as unlimited—many of them may well be.

If you're thinking of getting into the business, prepare for 6-8 months of planning, and then if equipment is available, set aside three to four weeks to test each cell site and conduct field tests. Finally, after spending about $3.5 million on all this, you can begin system tests that will tell you much you didn't know about the surrounding terrain and how your equipment should or could perform. Large systems, of course, cost more. By the way, of the 333 channels set aside for either wireline or non-wireline, you can count on only about 312 for general usage because of signaling set asides and adjacent channel interference. Like anything else that's highly technical and new, there will be a number of surprises awaiting even the best engineers and their entrepreneurs. Only well-financed, hardy pioneers really belong in the business. Successful returns certainly promise big bucks indeed!

Mobile Prices

In the beginning, mobile transceivers are expected to cost as much as $3,500. But by the second or third year, production and competition should reduce the price to the high teens and within five years to approximately $1,000. Prime U.S. producers are expected to be E.F. Johnson and Motorola. Offshore manufacturers known to be interested in the market are Oakie, Panasonic, Kokusai, and Hitachi. Oakie, by the way, supplied some units for initial Chicago tests long before the present systems were FCC authorized. These Japanese companies are said to have technical staffs equal to those in the U.S., and they may even offer transceivers ahead of most U.S. counterparts. What their serviceability factors will be, however, is unknown at the moment.

All U.S. types are expected to be plug-in and generally modular—an approach the Japanese have not adhered to in their offerings of consumer products. It's cheaper, of course to build nonmodular units and sometimes more reliable, but such units are considerably more difficult to service. For this and sophisticated automated checkout procedures, we'll just have to wait and see. Undoubtedly the marketplace will substantially influence this area. Sound and/or data reproduction and transmission, nonetheless, should be very much equivalent in the beginning before cost cutting becomes prevalent.

However, under a new program that's only an infant in 1983, the FCC will be monitoring more products than in past years, and those radio manufacturers whose quality falls below normal certification levels could, eventually, be in big trouble as the program matures. At least, this is the regulatory intent. What the final results will be several years down the road is anyone's guess. Much depends on the politics of the moment and the administration thereof.

Proposed Service and Quality

In the beginning, the mobiles may not offer many special features other than a telephone dial system, favorite numbers storage, and possibly short range wireless extensions through the cellphones. There may also be print material attachments to home computers to transfer data through the cellular system locally or elsewhere.

Surprisingly, because of FM constant-amplitude transmissions, special quieting circuits, 12 kHz bandpass, and excellent transceiver engineering cell phones will produce better voice reproduction and data transmission characteristics than standard phone service. In the Baltimore-Washington area, for instance, non-wireline American Radio Telephone Service, Inc. plans a system with 0.05 grade of service that ought to provide 95 percent call completion on the first attempt. This service will include 180,000 subscribers using 0.6 watt and 1 watt handhelds and mobiles within perimeters of 8 and 12 miles, respectively. Initially, American will service professionals, sales people, and marketeers. Movement towards recreational people and just plain consumers will occur as prices decrease. Apartment owners, for instance, would be naturals for this sort of service. Later, as rates change and conditions vary, it may become advantageous for some to switch to this service altogether. At this stage, who knows?

At any rate, all of these cellular systems are designed to serve a maximum number of subscribers; to make maximum use of available spectrum; to be instrument and frequency compatible; to provide regular and specialized service to mobile and land stations over both short and long distances; and to be available throughout much of the U.S. as well as abroad, if that's immediately possible, all at some reasonable cost. Some of these services will take a bit of doing in the near term, but down the road both distance and costs should be solved to some satisfactory degree.

Bell's Advanced Mobile Phone Service on wireline (AMPS) eventually perceives such advantages as Speed Calling (stored numbers available on instant recall); Call Waiting (signaling a busy phone that there's another call standing by on the line); and Three-Way Calling, where two parties may talk to a third party, switching back and forth between the connections. With all this going on, the Mobile Telephone Switching Office will record call durations, locations and destinations, time, originator, and all other pertinent parameters required for record keeping and billing. And at the end of the month Mr. Subscriber will receive a bill which could vary between a few dollars and hundreds of dollars, depending on both distance and usage. At the moment, about 20 cents/minute plus distance seems to be a going charge, but this will certainly vary with time, economic conditions, and competition.

Considering the AT&T operations divestiture in 1982-1983, cellular communications might become much more than another auxiliary service. It will be limited only by authorized spectrum and the service cells available. Ultimate cost could certainly become the determining factor. Consider the competition between the U.S. Post Office and the United Parcel Service. How do you ship your packages?

SYSTEM CONCEPTS

Whether wireline or non-wireline, the basic system setup is largely the same with the exception that the non-wireline groups may not necessarily connect directly to some long lines telephone network. This means that they would both use similar cellular concepts fed by the same types of mobiles and handheld units, but the non-wireline group would deliver far-field distribution either through fiber optics or microwave links. In the future, distribution could also be by satellite, especially in rural areas where neither microwave nor phone exchanges exist. NASA, the National Aeronautics and Space Administration, is already talking this up; but whether it becomes a reality with FCC blessing is another matter altogether. At the moment, such a possibility is rather remote and will probably remain so until the fledgling cellular industry is considerably more advanced than it is now.

Presently, the cellular spectrum has 40 MHz set aside for its use with 30 kHz assigned to each of 666 channels. This number includes 42 setup and signaling channels apportioned between the 825-845 MHz mobiles and the 870-890 MHz base stations in addition to a 20 MHz frequency reserve. These frequencies are to be shared by two systems in each of the 90 designated cities with certain rural areas thereafter assigned. So I am talking about 200-plus systems to cover the U.S. for starters, with the potential for added units in the unassigned 20 MHz area at some later date.

Cell Sites

Using a Bell System-type diagram, a wireline cellular system would look something like the one in Fig. 1-1. There you see the various cell sites grouped around a Mobile Telecommunications Switching Office (MTSO). Each of the cells, of course, receives and transmits messages to and from the mobiles and routes calls on to another cell site, another channel, or the MTSO, itself which would transfer such information into the ordinary telephone network.

A very similar type diagram, naturally, would apply to the non-wireline installation except that it might not connect directly with some telephone exchange. Either microwave or satellite uplink could easily take over the MTSO function, although a front office switch would have to change input/output frequencies and do housekeeping, such as billing, for the collective system.

In Fig. 1-1 you may have noticed that the cell sites were drawn as triangles rather than as some other type of figure. This was done deliberately to setup a contrast. In actual applications, circles aren't

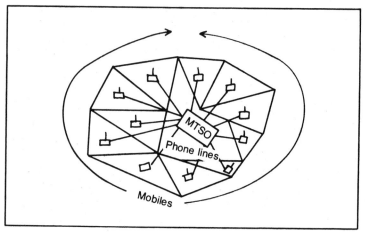

Fig. 1-1. Cell sites surrounded by mobiles, all feeding an MTSO switching office.

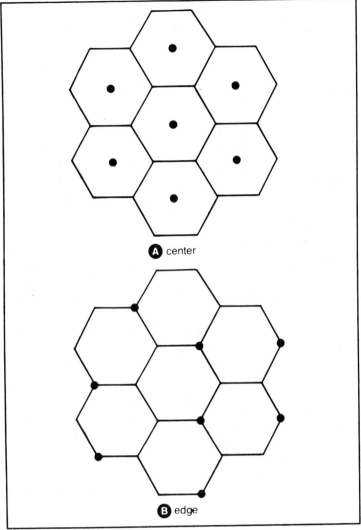

Fig. 1-2. Center (A) and edge (B) excited cells having omnidirectional antennas.

practical because of area gaps, while a square, triangle, or regular hexagon can completely cover any area without any lost spaces. So, to take advantage of a circular shape, but include the other geometric figures, a regular hexagon has been generally adopted as the most suitable configuration for each individual cell. Within these cells, of course, are the land transceivers on their cell sites. As Fig. 1-2 illustrates, antennas for the land stations may be omni-

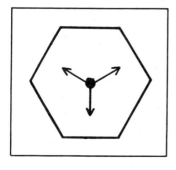

Fig. 1-3. Delta (120°) directional antennas for certain cells.

directional, either at each cell's center or on its corner. Another approach is seen in Fig. 1-3. Here a 120° separation is obtained with a Delta antenna configuration which is strictly directional rather than omni-directional as previously shown. Figure 1-4 shows a more detailed wireline cellular system. Note the three different ways to access the system.

Since cell sites are very expensive, only enough are constructed to cover the FCC-assigned area, and the least number of sites accompanied by minimum cochannel interference is a prime objective. Combined with the Mobile Telephone Switching Office (MTSO) and the little mobiles, the system is more or less complete. Frequency synthesized for 333 channels (wireline or otherwise), the control unit contains the microphone, earpiece, pushbuttons, and the various indicators while the logic unit uses its stored and reactive circuits to interpret mobile and land station calls and commands. When not calling or being called, the mobile samples the various channels and tunes to the strongest and responds to the data flow until there is a specific call that energizes certain responses. Figure 1-5 shows the block diagram of such a transceiver. The mobile remains on this channel unless signal strengths diminish, at which time it is automatically told to go to another channel and pick up the transferred transmissions there.

This transfer of channels is known as *handoff*, and it is specifically designed to maintain maximum signal communications at all times within the entire cellular system. And amazingly, the entire transfer is invisable to the user. Microprocessors handle the entire exchange. Cells remain effective as long as the transmitted signal-to-interference ratio does not become objectionable and bury incoming and outgoing information. Engineers have discovered that a cell's quarter radius measurement contains maximum signal but this decreases rapidly toward a cell's perimeter. Naturally, transmission power and receiver sensitivity also play a part in cell

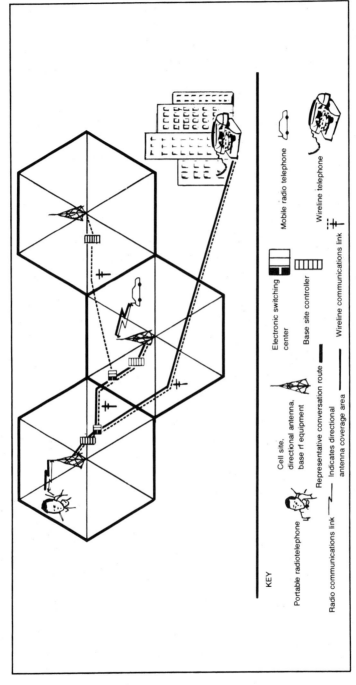

Fig. 1-4. A typical cellular system operation.

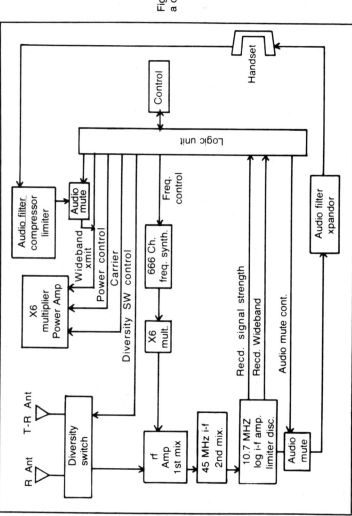

Fig. 1-5. Illustrative block diagram of a cellular mobile transceiver.

efficiency. Units with prime responses and maximum meantime between breakdowns contribute greatly to subscriber satisfaction and minimal cost of cell site operation.

Several years ago, 10 watts was considered the magic number for transmitters in the 800-900 MHz spectrum, requiring some 12 watts from mobiles and about 40 watts from the cell sites because of combiner, cable, and other losses. At that time, Bell engineers estimated antenna gain to be between 6-8 dB relative to some dipole and the antenna elevation at about 100 to 200 feet. As we proceed you'll note some considerable changes—although under certain terrain conditions, this much power and height may be required to do the job.

They also established a signal-to-interference ratio (S/I) of 17 dB or better over 90 percent of system coverage. However, with a 30 kHz channel spacing and peak signal deviation of 12 kHz divided into 312 effective channels (624, in all), only a few similar channels belong to any particular channel set, and, therefore, the probability of adjacent channel interference is small. Directional antennas and good front-to-back ratios also avoid considerable interference problems likely to occur. Sequentially numbered cells may well be assigned adjacent channels, and these can be oriented so they point in different directions, reducing further possible interference problems.

If there is cell splitting, as shown in Fig. 1-6, the distance between adjacent sites is halved, but the density is quadrupled,

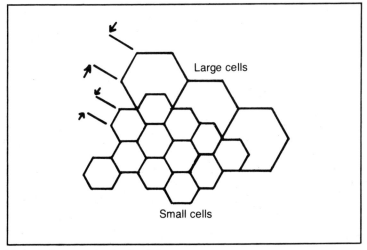

Fig. 1-6. An example of cell splitting.

according to Bell. And wherever dual-size cells reside, channel numbers must be rearranged accordingly—a warning that an initial system should be laid out carefully in the beginning to avoid costly complications later on. Of course, when systems are started in non-densely populated areas, such cell-splitting arrangements may become necessary to handle the traffic as demand and users grow. Nonetheless, a little foresight and careful planning could eliminate many a headache in the future, especially in difficult terrain or with complex spectrum usage.

Home Mobiles and Roamers

Subscriber mobiles within the Standard Metropolitan Statistical Areas (SMSAs) and those without are colloquially called *Home Mobiles* and *Roamers,* respectively. AMPS and the independents need to offer service both to those within the SMSA and to the Roamers as well as to distant contacts with whom local subscribers wish to communicate. Here, the standard 10-digit local and distant telephone number permits interaction between the phone network and the callers, wherever they may be. The analog number is first dialed, then stored, and finally transmitted very rapidly in digital form. The subscriber's control identification and serial numbers permit the call to continue on its way via the assigned cell site and switching office or go directly from the cell site to another Home Mobile or Roamer, depending on circumstances.

At the same time, the originating equipment and the cell site continue to exchange housekeeping information to maintain adequate open channels and synchronization with the Mobile Telecommunications Switching Office (or its independent equivalent) monitoring calling/receiving information and billing data.

Supervisory signals in AMPS include tone burst and continuous out-of-band modulation known as ST signaling tone and SAT supervisory audio tones of 5790, 6000, and 6030 Hz. At any given cell site, only one of these tones is used, and a closed loop becomes established between the mobile and the cell site which only recognizes its particular SAT frequency. Either of the other two are interpreted as rf interference. In mobile-to-cell communications, a 10 kHz ST tone appears upon user alert, handoff, certain service, or disconnect. Locating and handoffs continue to maintain mobile signal strength at usable levels automatically regardless of where they may be operating within the system. MTSO facilities routinely monitor both gross range and rf signals as the mobiles pursue their travels.

Paging and Access

In cellular communications, *paging* of the mobile means discovering if it can take a call; *access* is simply receiving the call. Here, the 21 signaling and setup channels in each wireline and non-wireline setup are used since hundreds of channels may be involved in the process, especially when there are handoffs and other extenuating circumstances. And as the number of subscribers grow so does the channel capacity. That's the prime difficulty with paging. As for access, rapid and accurate response from the mobiles is very often proportional to the number of subscribers. The mobile must respond to an inquiry, furnish identification and numbers, and wait for an assigned channel. In the event of overload, the mobile has to recover rapidly and offer correct responses whenever its numbers are called.

Conversely, paging must find the proper mobile, supply enough channels for at least continental U.S. mobile, and roaming and be sufficiently expandable for future requirements. When you think of it, those paging and access demands are pretty tall orders for any wireline and/or over-the-air system to satisfactorily service. What happens when hundreds of cellular units dot the cities and countrysides remains to be seen; but you can bet there will be at least some confusion under certain circumstances which are now probably unforseen. The cellular concept is obviously fascinating, but like every new system it's going to have some bugs. Both subscribers and engineering will find out soon enough as service areas spread.

Receiving and Answering Process

When any mobile is in operation, it scans assigned and dedicated control channels in either System A or System B, depending on whether the serving-system status is enabled or disabled. Channel signal strengths are automatically examined, and the mobile is tuned to the strongest channel where it waits for an overhead message train from the cell site to verify its system identification, roam status, and local control condition. All this should be completed on either the first or second strongest paging channel. If, in a period of three seconds, there's no verification, the serving system status must be rechecked and the mobile has to go to scan-dedicated control channels.

When the mobile responds to an overhead message as in Fig. 1-7 it must update its serial number, station class, registration bit, extended addresss, discontinuous transmission, paging channels, read-control filler, and access channels, local control, and ID for

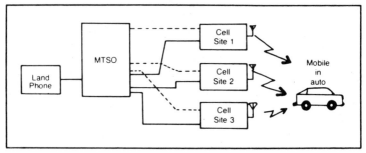

Fig. 1-7. Cell site selection.

autonomous registration. The serial number has 32 bits, autonomous registration has 4 bits, next registration has 21 bits and the system ID has 15 bits. The mobile must also have an 11-bit first paging channel number. Further, the supervisory audio tone (SAT) should be identified within each 250 milliseconds (that's the group of 5970/6000/6030 Hz tone(s) modulated on the transmitted voice channel carrier). This is quite a bit of fast-response housekeeping just to say, "I'm ready to receive and transmit the land stations wishes." Maximum information bit rate between the two amounts to about 1200 bits per second for information throughput. This bit rate is doubled between the cell site and the Mobile Telephone Switching Office (MTSO) in the AMPS system.

After data transfer or voice conversation, either the land telephone or the mobile telephone can disconnect, usually by simply hanging up the receiver. The mobile then usually emits a signaling tone and shuts down its transmitter, while MTSO goes into idle, issuing whatever disconnects are required. Finally, the cell site also shuts down its transmitter to await further service. According to Bell, a large mobile system could have some 50 cell sites, 100,000 subscribers, and a single MTSO. Undoubtedly these figures will change as actual systems assume operational status in late 1983 or 1984.

Mobile Telephone Switching Office

In AMPS (Bell Telephone) language, MTSO stands for Mobile Telephone Switching Office, and it is the name given to its central coordinating element. As mobiles and land communications proceed, the MTSO receives all AMPS mobile communications over a large area, if not the complete system. Designated as common control, the MTSO is made up of signal processors, memories, switching networks, trunk circuits, and ancillary services. The processors and the memory are redundant. Switching networks interconnect the cell site trunks and the phone network. The programs stored in the switch memory control MTSO operations. The cell sites offer voice communications via physical connection to the MTSO through the trunks. There are also two 9.6 kilobit/second data links connected between the cell sites and the MTSO that operate in full duplex. Should one data link fail, the other active link can take over.

The MTSO, of course, processes all mobile phone calls under programmed control and does the following:

- ☐ Computes time and billing information.
- ☐ Offers the mobile subscriber custom servies.
- ☐ Controls handoffs.
- ☐ Furnishes switched interconnects with land phones.
- ☐ Switches for mobile subscribers using MTSO.

Mobile Operations. As mobiles call into MTSO, each message contains the mobile's ID, the complete number called, and the serving-cell identification. Incorrect or incomplete messages are rejected and reordered after the MTSO orders a designated voice channel setup. Where successful, the MTSO seizes a network trunk and outpulses the particular number. Cell site trunks and outgoing trunks are now switch-connected and a talking channel is established. Figure 1-8 illustrates this idea.

Messages on incoming trunks (a call to a mobile) are first collected as digits, analyzed and identified as being proper or improper for the called subscriber, then either processed as a ringing tone to the caller or routed as an invalid number to be intercepted. On valid numbers, the MTSO begins systemwide paging to the various paging cell sites to locate the mobile.

If no response is forthcoming, even on the second try, the caller is automatically connected to a recording announcing failure.

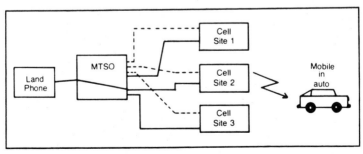

Fig. 1-8. Conversations under way.

But when a page response does appear, an idle trunk is seized; the cell site, instructed by the MTSO, sets up on the selected voice channel. As overhead communications become linked, the trunk and cell sites connect, thus establishing throughput.

Once conversations begin, occupied channels are cell supervised for adequate mobile signal quality and strength. When inadequate conditions occur, an automatic handoff to another voice channel can be effected, or the call is left undisturbed because another channel cannot help and conversation continues or not as the mobile moves further in or further out of range. Handoffs conducted by the MTSO proceed as follows:

☐ A trunk is designated for the new cell site.

☐ Both cell sites must execute for the new voice channel.

☐ The talking path is channeled from the new cell site trunk into the outgoing trunk in a ready state.

☐ The talking path within the switching network from old cell site to outgoing trunk is idled.

As a handoff takes place, a new cell site trunk and idle voice channel are designated by MTSO, and the new cell site must turn on the proper radio. As this occurs, a handoff order message goes to the serving cell site containing the new voice channel identity. The MTSO then waits for a message from the serving cell that mobile communications are now on the new voice channel, the new talking trunk path becomes reconfigured, and the handoff is complete.

When the conversation is completed, a disconnect takes place to free the channel for others. Should the mobile disconnect first, a 10 kHz signaling tone ensues for the cell site, and the mobile turns off its transmitter. The cell site then shuts down the serving voice channel transmitter and releases the channel to the MTSO, resulting in an idling of channel control. Should the land or base station exit first, MTSO instructs the serving cell to release the call. Confirmation is awaited, and if release is effected, the channel is idled. In the same way, if calls are lost because of signal failure at the mobile or at the cell site locations, the call will also be properly terminated.

Should a SAT (Supervisory Audio Tone) not be heard by the cell site within 5-6 seconds, it assumes call termination and notifies MTSO of an abnormal release. This is so recorded, billing changes made, and the channel is idled. When a mobile loses SAT for this same period, its transmitter turns off, setup channels are scanned, and the strongest channel is selected. Cell sites which have not noted the SAT loss do so when a mobile terminates. Unsuccessful calls may also be problems and result from such abnormalities as signaling and dialing errors, busy lines, equipment failure, and traffic blocks. Recording announcements and tones identify call failures to the originator. Tones may also originate from the mobile, but MTSO controls application of these tones whenever they are used.

For instance, if a mobile is on a voice channel, the MTSO connects the cell-site trunk and tone source; but if there is a problem before voice channel connection, then the MTSO tells the mobile to produce an internal tone. The 10 kHz signaling tone (ST), on the other hand, is for user alert, for a handoff, for disconnect, or for flashing as for a hold. SAT tones are 5970, 6000, and 6030 Hz with only one being used per cell site. This tone is transmitted to a mobile which returns the tone and closes the power loop.

Billing. This task is one of the prime duties of the MTSO. All charges with accompanying information are recorded on tape, including a record of calls to and from all mobiles, with entries noted on those that accessed a voice channel. Data includes phone numbers, toll charges, channel seizure and release, cell identification, and answer and disconnect times. For completed calls, the call-originating office supplies billing relative to the use of the land phone network.

Housekeeping. The MTSO also accounts for the number of data links, cell sites, trunks, service circuits, switching networks, and internal memory. According to AMPS, counts are made of total seizures, blocked attempts, and overall usage. The latter is sampled at intervals of 10 and 100 seconds. Usage tells engineers the number of various equipments that are busy during the sampling times so that the system may activate sufficiently to keep the traffic moving satisfactorily. These measurements are seen on MTSO printouts several times during each day.

AN AMPS CELL

In the foregoing cell discussion, I primarily covered the subject generally, and not specifically from the wireline standpoint. Here, I'll try and go into more telephone-oriented detail from material generously supplied by AMPS from its New Jersey headquarters.

As you have already discovered, any cell (or cell site) connects land and mobile units, and the AMPS cell specifically connects radio to the various channels and their trunk terminations. According to AMPS, such cell sites have the following duties:

☐ Provide call setup, supervision, and terminations.
☐ Handle mobile controls during calls.

- ☐ Send, receive, and distribute calls.
- ☐ Channel data between the mobiles and MTSO.
- ☐ Locate mobiles.
- ☐ Do voice processing.
- ☐ Execute equipment control, maintenance testing, and fault detection, in addition to certain repairs.

Since these cells and their sites stand alone and are largely unattended, the foregoing functions are necessary to continued an uninterrupted service. Figure 1-9 provides an overall block diagram.

As you may recall, cellular mobile/telephone systems can handle many more simultaneous calls than those which might be accommodated on individual channel frequencies. Therefore, cell operation, its directional or nondirectional antennas and responsive electronics play a major part in total cellular communications. Selective cell malfunctions could easily interrupt an entire system if these were not constantly monitored and promptly serviced. Most AMPS systems are scheduled to operate with a 7-cell repeat pattern, although as many as 12 and some less than 7 may be installed, depending on system growth and propagation requirements.

Cell Makeup

Any AMPS cell site is electronically controlled by a microprocessor Module Control Unit (MCU) which is divided into two identical halves. One is active and the other passive. If the active half experiences trouble, the inactive half is activated, duplicating all prior services. Redundancy also extends to other critical subsystems, making the system as reliable as possible. There is even battery backup in the event of commercial power loss. As you might surmise, most of this equipment is modular and may be replaced or added to handle voice and data traffic as needed.

Equipment comes in the form of *frames* consisting of major elements called Radio Frames and Control Frames (see Fig. 1-9). Any cell site may have as many as 6 Radio Frames containing a maximum of 96 voice radio channels. One Control Frame handles each cell site and furnishes voice and data mobile links with MTSO and repair-maintenance operations. Voice channels may also be programmed to handle information other than voice, and this is done by a technique known as *blank and burst*. This means that voice signals are blanked and data is then rapidly transmitted at nearly maximum bandwidth consumption in the form of burst at a rate of some 2400 bits per second. Similar to a botched call on wirelines,

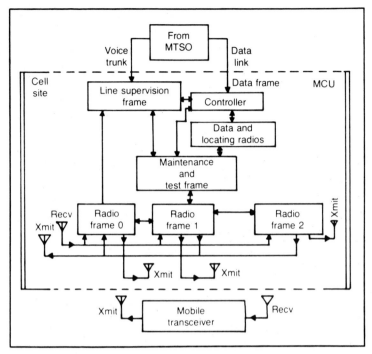

Fig. 1-9. AMPS cell with MTSO and mobile connections.

the *falsing*, or failure bit rate, of such channels may be relaxed to 10^{-5}. In voice channel data formats, messages are repeated 11 times forward, but only 5 times in reverse. AMPS says that handoff messages, the critical portion because of greatest possiblility for errors, are usually returned under low signal-to-interference (S/I) conditions and, therefore, have a low completion rate. Coding amounts to 32,26 BCH with sync contained within the start of each message. Initially there is paging, followed by cell site selection, reply, channel assignment, signal alert, and communications for the standard mobile sequence.

Blank and burst signaling often have to be used in orders such as handoffs and alerts to mobiles. They take the form of binary data, require only about 1/10 of a second, and are not normally noticed by the subscriber. Initiated by either cell sites or the MTSO, blank and bursts can go in a forward direction, and reverse messages can be transmitted by mobiles. The mobile first sends 10 kHz of signaling tone to the cell site, and receives a reply from the cell site which has switched in a data decoder. By reverse blank and burst, the mobile transmits its message which may either be in response to a forward blank and burst or by subscriber request for some special service.

Setup Radios, used only in data transmissions, actually do the *make ready* work before voice communications are established. They share in cell use for mobile communications, as well as transmit overhead messages to mobiles with cell access which are momentarily idle. As usual, the mobile supplies forward and reverse setup and voice channels to the cell site while the cell furnishes voice circuits and shares a 9600 bits/second data link with the MTSO. From land-to-mobile, such radio messages may either be one or two words long, each consisting of serialized data bits which contain both messages and error detection and correction bits. From mobile-to-land, similar words and content are transmitted, but the words may be more or less, and such words appear as part of the message. Each mobile only looks at one of the two message stream words, depending on whether its final phone digit is odd or even.

Forward setup channels have special busy-idle bits. If the busy-idle bit is 1, the reverse setup channel of some cell transmitting is idle; but if it is a 0, the reverse setup channel is in use and its access will have to wait. Cells have between two and four setup radios placed in the control frame. Module control units also have *Location Radios* that measure signal strengths and other characteristics of the received signals. New voice channel selections are processed on the basis of such measurements since these radios may be tuned over the entire received band. There are two Location Radios for each cell site.

Talking Channels

Before there is radio baseband transmit and receive with the phone network, specified control actions are added in transmit and others deleted from receive. Later a syllabic compandor compresses speech variations in transmission and expands them in reception. Gains and losses must complement one another so that final signal levels appear unaffected. In ordinary FM radio, this general action is called preemphasis and deemphasis, permitting higher frequencies to be emphasized from the sending end and deemphasized at the receiver, although there is no actual operation on speech power levels as there is here. Mobile as well as cell site equipment contain these compandors.

The Transceiver

Each audio channel has a transceiver module phase modulated

by voice and SAT or frequency modulated at the 10 kbits/sec data rate. Channel frequencies are synthesized and generated upon command from a 10-bit control word and initial power output amounts to 1 watt, which is later boosted to a maximum of 40-100 watts by the final power amplifier. A channel multiplexer combines all carriers (a maximum of 16 per cell) onto a coax feed line connected to the transmit antenna. The usual 45 MHz spacing separates transmit and receive operations, and channel spacing remains at 30 kHz.

In receive, these AMPS units pick up signals from a pair of antennas delivering inputs to broadband amplifiers and power splitters. Baseband voice, SAT, or data signals are demodulated (detected) and delivered to the system. Like the transmitter, the receiver is also frequency synthesized, deriving its timing from the same oscillator.

It's interesting to note that in a Bell Telephone study, received signals have delay spreads of as much as 3.5 μsec in urban areas. Fortunately, the delay spread-related coherence bandwidth with fading as a 0.9 or greater correlation will not significantly affect transmissions within the assigned 30 kHz bandwidth. Bandwidths which would be affected, according to Bell, would be greater than 40 kHz in urban and greater than 250 kHz in suburban areas. Fade durations are inversely proportional to vehicular speeds, so the faster you go the less the fade; but do remain within normal speed limits.

Antennas

Antennas and propagation losses are very important subjects in cellular systems since communications depend on both antenna gain and location as well as natural and manmade obstacles in between. Ground planes in shape and size are also factors, especially when mounted on moving vehicles. So there's a good deal more to dealing with the 800-900 MHz band than the old land mobile in the 25-50 MHz region of yore. Further, antenna gain has to be measured as a function of the angle of elevation taken by its signal path, and urban/rural responses differ greatly because of unpredictable terrain. Reflective signal paths are another hazard, and antennas that measure gain under laboratory conditions may offer little extra signal reception/radiation when mobile. Whip antennas with gain may also deliver little or no improvement to the fading problem. So, depending on specific conditions at the time of both use and measurement, the choice of some mobile antenna may or may not be of considerable moment, but its surroundings and ground plane can certainly make a difference. This is why several new types of antennas are expected to appear as cellular radios become more abundant and experience teaches with use. And it may be that some worthwhile use can be made of multipath reception, if it can be corrected in phase.

Victor Graziano, of Motorola, Schaumburg, Ill. has produced a very worthwhile paper on propagation models for 900 MHz cellular systems using portables and mobiles. In it he deals with mean path losses for street coverage in rural, suburban, and urban areas. Some of the study involves ground and nonground reflections, horizon distances and earth curvatures, uneven terrain, both natural and manmade. He reports that "subjective tests of portable operation at 900 MHz in the Washington-Baltimore area gave results in reasonable agreement with predictions of 2 to 5 miles obtained using an inbuilding portable coverage model." Here, American Radio Telephone Service and the Washington Post have now had their cellular non-wireline service approved for review of full service operation by the FCC on April 27, 1983.

This system, as described by Arnold Brenner, V.P. and Director of International Product Operations, Motorola, now has 36 voice channels, 7 signaling channels and 1 EMX. There are six 60° sector receive antennas, and the transmit antennas with 40 watts input, produce a 100 W output ERP. In the handoff operation, the system determines bearing and distance from the base or cell site. Each of the six directional antennas and their receiver measures the mobile's carrier power, and the one with the strongest signal then records any phase delay of the 6 kHz tone. Bearing and distance is then stored in the base site controller and continually updated. Also, a maximal ratio combiner operates between adjacent cells to combine signals resulting from multipath propagation.

With antennas and receiver continually updating incoming data, cell-to-cell handoffs also proceed smoothly if the new cell has a channel available. If not, the transmission is placed on high priority queue and regains cellular handling when another channel opens up. Handoffs require only 50 milliseconds to complete. Typical service areas furnish 90 percent of the subscribers with better than minimum signal strength and at least 17 dB C/I ratio, offering excellent voice quality. Cellular, however, is not designed for dispatching service since message lengths are normally only a few seconds, and repeater systems with either satellite receivers or simulcast would be much more economical to implement than a cellular system.

Transmit and receive antennas in smaller or initial cellular systems are normally high gain, vertically polarized and omnidirectional. Gains of 9 dB are not unusual. Figure 1-10 illustrates such an antenna.

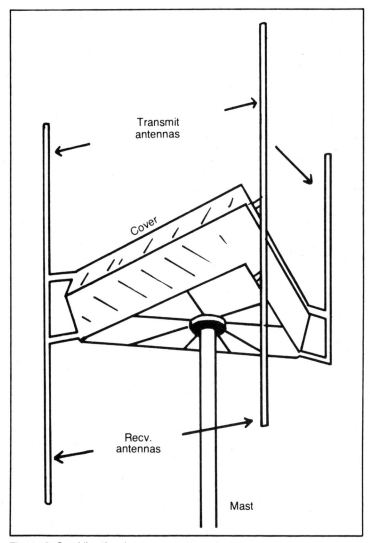

Fig. 1-10. Omnidirectional antennas and receive antennas mounted on mast.

A single transmit antenna handles 8-16 radio channels, while dual receive antennas supply signals to the 2-branch space diversity input. Both are centerfed, measure some 13 feet in length including mounting, and are protected by a 2.5-inch fiberglass covering. Where there is cell splitting requiring directional antennas, they may either be mounted on freestanding masts or on existing structures immediately available.

Receive spatial diversity antennas actually co-phase their incoming signals which are then added before baseband detection. This normally improves reception since there is small probability that antennas spaced a half wavelength apart will be subject to identical signal fades. Of course, the cost is higher both in antennas and receiving equipment for this sophisticated type of reception, but improvement in voice signal-to-noise ratios under less than desirable conditions is substantial, often in excess of 10 dB.

This subject does, indeed, merit more attention than these few lines and will be considerably more extensively treated later in the text where additional information is to be given. Antennas and transmission lines, unfortunately, are subjects too many authors avoid because their concepts are somewhat difficult if there is no familiarity. Nevertheless, there is no way that 2-way radio can operate without efficient transducers and their coaxial conduits. Impedances, losses, lengths, standing wave ratios, front-to-back-ratios, and signal attenuation all decidedly enter into investigative calculations and final observed results. A poor radio may even do well with a good transducer; but a good radio makes little difference with poor quality or defective radiator and coax. Signals must get in and get out or they will never be sensed or heard regardless of transceiver quality. In any type of communications, therefore, you begin with good antennas and adequate cabling, then proceed to lesser or greater sophistication in the sending and receiving apparatus. Quality products usually last longer and perform better than their less adequate counterparts.

Obviously, however, as production and product familiarity increase, prices will drop and quality reception and transmission advance. In the meantime, there will be enormous amounts of money spent in erecting and testing these cellular systems as fast as the Federal Communications Commission gives its various system approvals. And dozens of new systems will have that approval even before this book is published.

Chapter 2

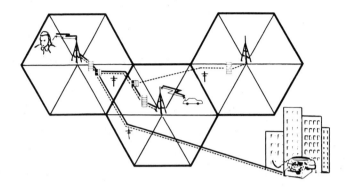

Terrestrial and Satellite Carriers

This chapter is divided into wireline, non-wireline, and satellite carriers who have gained some initial approval or are on their way to further consideration by the FCC. These carriers are Advanced Mobile Phone Service, Inc., Western Union, and the National Aeronautics & Space Administration.

Similarities between the wireline and non-wireline systems are considerable in many respects except that control points in the former are not radio connected and there's no multiplex. However, cell sites, mobiles, hand held units, and main switching offices for recording and phone connections are all there in both operations, and each will cover similar territories since one wireline and one non-wireline system is authorized for the top 90 U.S. cities. Thereafter, NASA hopes to take voice and data communications topside into one or more geosynchronous satellites from rural subscribers and spread the word back to earth at appropriate locations across North America—at least to begin with. At the moment, my predictions tell little of what the future holds for this remarkable medium since the medium itself has hardly begun to have growing pains following initial construction on the heels of two or three experimental units around the country.

I am well aware the system works. Otherwise the FCC wouldn't have approved it. But how well it works and how many subscribers are to pay the monthly charges is something else again.

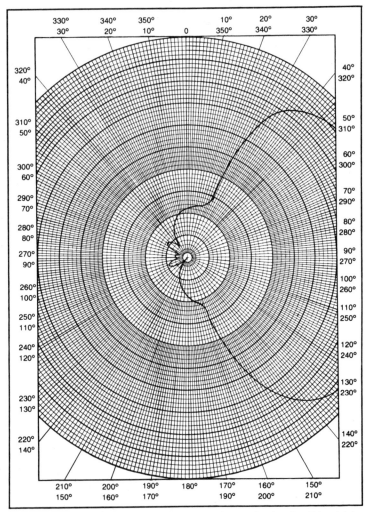

Fig. 2-1. Horizontal beamwidth, directional and front-to-back ratio of Mundelein, Ill. cellular land station antenna. (courtesy AMPS)

Projections are relatively lavish, but another several years will determine the proof.

AMPS IN CHICAGO

Prime wireline cellular service AMPS (Advanced Mobile Phone Service, Inc.), a wholly owned subsidiary of AT&T, has already been approved by the FCC to operate in the Chicago, Illinois area, among others.

It is authorized to construct and test a "Domestic Public Cellular Radio Telecommunications System, consisting of 17 base stations operating on cellular System B frequencies." The lead application for Base Station C, upon which others are to be patterned, originates from Mundelein, Illinois, east of the Kruckenberg and Fairfield Roads intersection. A total of 75,000 mobiles for the Chicago area are anticipated.

System Design and Configuration

For the Chicago cellular grouping, the design specifies a grid of hexagonal cells collected into repeating groups of seven cells each, whose numbers are limited by cochannel interference possibilities. AMPS says that if R is the cell radius and D the distance between neighboring cochannel cell centers, then

$$\frac{D}{R} = \sqrt{3N}$$

And if the $\frac{D}{R}$ ratio offers acceptable system operational quality, then squaring both sides gives the following:

$$21.16 = 3N$$
$$N = \frac{21.16}{3} \text{ or } 7.5$$

N establishes the desirable number of cells per group (N). Those not separated by normal reuse distance must be assigned different operating channels.

Reusable frequencies such as F1, according to AMPS in Fig. 2-2, may be placed in other nearby F1 cells as indicated. Figure 2-3, illustrates control and voice frequencies for Block "B" base station transmit assignments for the first seven channels. Note that control is always 630 kHz less than the voice channel below. To convert base station transmissions to mobile transmissions, multiply the channel number times 0.03 and then add 825 MHz. For base station transmit frequencies, multiply the channel number times 0.03 and add 870 MHz.

The Bell system, in technical reference PUB 43303, specifies the AMPS and Public Switched Telecommunications Network interconnect. Each connecting circuit delivers one 4-wire, voice grade, E & M signaling interface to the MTSO with 2-way trunk operation, seizure, answer, disconnect supervisory signals, and address. The MTSO and the cell sites also have 4-wire connections called radio land lines. One radio land line is required for each cell

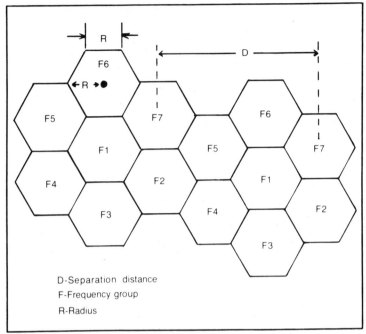

Fig. 2-2. Cell map illustrating reusable frequency grouping. (courtesy AMPS)

site channel and for two full duplex data channels.

Ten Western Electric 40F9 voice transmitters and two 40F9Y control WECO transmitters are to be used in the system. The control point, located in Hickory Hills, is to have an antenna with overall height of 117 feet above ground. This figure translates to 992 feet above sea level. Radio is not to be used to connect control points and there is no multiplex contemplated. Antenna power gain, referenced to a ½-wave dipole, amounts to 9 dB. The 12 CMP-1 WECO transmitters are to have a frequency stability of 0.00015 percent, and communications test sets with ±0.00005 percent accuracy are to be used for frequency measurements. Estimated cost (presumably) for the Mundelein facility is listed as $520,738. AMPS says each base station will have test facilities and measuring instruments for diagnostics tests on either channels or base station equipment, including the following: antenna output power, transmitter frequency and deviation, and receiver sensitivity. Routine automatic tests are also expected. The horizontal pattern of Mundelein's directional antenna in the horizontal plane is shown in Fig. 2-1. Spelled out, you see a beamwidth at 3 dB down of 105° (between 35° and 140°) and a front-to-back ratio of approximately

	Channel Set 1		Channel Set 2		Channel Set 3		Channel Set 4		Channel Set 5		Channel Set 6		Channel Set 7	
	Chan. No.	Base Freq.	Chan. No.	Base Freq.	Chan. No.	Base Freq.	Chan. No.	Base Freq.	Chan. No.	Base Freq.	Chan. No.	Base Freq.	Chan. No.	Base Freq.
Control	(334)	880.020	(335)	880.050	(336)	880.080	(337)	880.110	(338)	880.140	(339)	880.170	(340)	880.200
Voice	(355)	880.650	(356)	880.680	(357)	880.710	(358)	880.740	(359)	880.770	(360)	880.800	(361)	880.830
	(376)	881.280	(377)	881.310	(378)	881.340	(379)	881.370	(380)	881.400	(381)	881.430	(382)	881.460
	(397)	881.910	(398)	881.940	(399)	881.970	(400)	882.000	(401)	882.030	(402)	882.060	(403)	882.090
	(418)	882.540	(419)	882.570	(420)	882.600	(421)	882.630	(422)	882.660	(423)	882.690	(424)	882.720
	(439)	883.170	(440)	883.200	(441)	883.230	(442)	883.260	(443)	883.290	(444)	883.320	(445)	883.350
	(460)	883.800	(461)	883.830	(462)	883.860	(463)	883.890	(464)	883.920	(465)	883.950	(466)	883.980
	(481)	884.430	(482)	884.460	(483)	884.490	(484)	884.520	(485)	884.550	(486)	884.580	(487)	884.610
	(502)	885.060	(503)	885.090	(504)	885.120	(505)	885.150	(506)	885.180	(507)	885.210	(508)	885.240
	(523)	885.690	(524)	885.720	(525)	885.750	(526)	885.780	(527)	885.810	(528)	885.840	(529)	885.870
	(544)	886.320	(545)	886.350	(546)	886.380	(547)	886.410	(548)	886.440	(549)	886.470	(550)	886.500
	(565)	886.950	(566)	886.980	(567)	887.010	(568)	887.040	(569)	887.070	(570)	887.100	(571)	887.130
	(586)	887.580	(587)	887.610	(588)	887.640	(589)	887.670	(590)	887.700	(591)	887.730	(592)	887.760
	(607)	888.210	(608)	888.240	(609)	888.270	(610)	888.300	(611)	888.330	(612)	888.360	(613)	888.390
	(628)	888.840	(629)	888.870	(630)	888.900	(631)	888.930	(632)	888.960	(633)	888.990	(634)	889.020
	(649)	889.470	(650)	889.500	(651)	889.530	(652)	889.560	(653)	889.590	(654)	889.620	(655)	889.650

Fig. 2-3. Chicago AMPS cellular system frequency block B channel assignments. (courtesy AMPS)

22:1 with no problem sidelobes and inconsequential backlobes. This, of course, is the 9 dB gain antenna referenced to a ½ wave dipole mentioned earlier. Voice transmission line data shows a +12.4 dBW gain out of the CMP-1 transmitter, a loss of −1.4 dB for 125-feet of line, and a 9 dB antenna gain using Andrews HJ5-50A feedline into an omnidirectional antenna with a total effective radiated power (ERP) of 100 watts. The control channel information supplies a +9.4 dBW CMP-1 transmitter output, a loss of −0.8 dB at 125-feet, and a +9 dB gain for the omnidirectional antenna.

Total construction costs as estimated by AMPS come to $18.898 million. This includes $9.413 million for cell sites, $5.527 million for the MTSO, $0.448 million for prorated ACC costs, and $3.510 million for other costs. Operating expenses for the first year were estimated at $4.655 million. As you can see, 17 base and/or control stations cost a bundle. On the contrary, $160 to $185 per month on the average to the reseller times at least 70,000 mobiles provides almost $12 million/month of basic income when in full operation. Quite a tidy sum if operating expenses are kept within reasonable limits and subscriber use lives up to expectations. This is why mobile cellular appeals to such a large group of entrepreneurs, both wireline and non-wireline. However, as AMPS estimates, the break even point for the entire system numbers some 4200 subscribers offering a total revenue of $7.4 million per month. So there is a great deal here that won't appear as pure profit, and there are always hidden expenses that arise from time to time and pare the returns even more. One subscriber may operate more than one mobile, so the $12 million/month initial figure may not precisely apply.

Base Stations

AMPS Chicago has adopted an alphabetical means of identifying primary, secondary, and tertiary base station locations. And for system identification CHI will stand for Chicago. The nomenclature method is as follows.

☐ Primary locations: A - Z and AA - ZZ.

☐ Secondary locations: These are halfway between the primary locations and so are identified as A,B if between A and B.

☐ Tertiary (third) locations: Are halfway between two secondaries or one primary and one secondary. A,B and A,C secondaries, as an example, would become the tertiary AB, AC while designation for primary location A and secondary location A,B, identify as A, AB, all of which are slight reminders of Boolean logic, the Law of

Duality, and de Morgan's theorem; all begun by George Boole of Dublin in 1854. Remember?

☐ Theoretical locations: To find certain base station locations, use ZZ and "descend alphabetically" for as many theoretical locations needed to find the actual station. A base station located between primary A and ZZ would become secondary A,ZZ.

AMPS says all base stations will be connected to one another through a common switching arrangement identified as the Mobile Telephone Switching Office (MTSO), with radios interconnected with the public landline phone network. Any given set of frequencies may be reused in nonadjacent cells or another part of the system, accompanied by handoffs for adequate signal levels. The system is also designed to evolve into smaller cells that will handle the large number of expected subscribers.

Calling Actions

An MTSO receives, stores, and verifies dialed *mobile calls*, and then occupies a central office connecting circuit. When the central office winks a start signal, the MTSO transfers calling numbers to the central office which sends them on, meanwhile relaying an answering signal to MTSO. These connections are shown in Fig. 2-4. When the call is completed, the MTSO accounts for the call and bills it to an appropriate address. Although AMPS does not offer its own operator services, regular phone company operators are available upon dialing 0,0+ called number, but such assistance will be limited to credit cards, collections, and third number billing.

Land calls to mobiles are based on phone numbers assigned from a central interconnect office. An idle central office circuit to the MTSO is seized, and, when a start wink from MTSO occurs, the central office transmits (outpulses) the usual four or five digits needed to identify the mobile. When the call is completed, an answer signal goes to the Network and disconnect commands are sent to and from the Network to shut down connections.

Mobile-mobile calls within the local system are completed through MTSO and Network intervention is not required.

Frequency Allocation Durations

Following Federal Register Publication sometime after February 25, 1982, non-wireline carriers and wireline carriers are to apply for their respectively assigned Block A and Block B set of

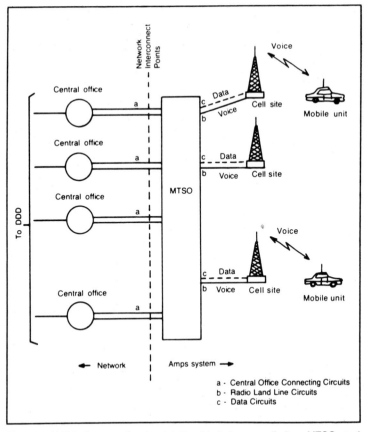

Fig. 2-4. AMPS diagram showing relationship between cell sites, MTSO, and central office. (courtesy AMPS)

frequencies. But following a two-year period from the Federal Register date, the FCC will permit anyone to apply for "either frequency block in any market, except where the block has been applied for during the term of the separate allocation." Previously, the FCC had stipulated a period of five years for the two types of carriers to remain separate. This period should have vanished by March of 1984. The FCC also deleted any requirement that wireline carriers have to offer cellular service through a separate subsidiary, except to the extent it applies to AT&T. But if AT&T or other affiliates can show unique factors where benefits are outweighed by certain costs, then the FCC will consider waivers, but does not necessarily guarantee favorable action. Meanwhile, although applicants may propose base station equipment for systems not yet type accepted, both base stations and mobiles have to be Federally approved before their entry into service.

WESTERN UNION IN BUFFALO

The first non-wireline cellular system to gain approval from the FCC is also in Buffalo, consisting of a consortium headed by Western Union.

Western Union has also purchased a substantial minority interest in Cellular Communications, Inc. of New York. This company is expected to be at least one of the cellular mobile radio telephone systems construction arms for Western Union in many metropolitan areas throughout the country. Obviously indicating its very serious approach to cellular communications, Western Union Telegraph Co. and six joint venture partners have already filed applications with the Federal Communications Commission for authority to construct and operate cellular mobile radio telephone systems in 27 of the 30 second round cellular markets. WUTC has also filed for licenses in Birmingham, Ala., and Richmond, Va., and in 25 other markets with other companies or where it has equity interests. Cellular Communications, Inc., is an example, and has filed for Columbus, Sacramento, Memphis, Dayton, Bridgeport, Norfolk, New Haven, Akron, and Orlando. FMI Financial Corp., another joint venturer with WUTC is said to have filed with the FCC for cites in Louisville, Nashville, Jacksonville, and the Westel Companies for Indianapolis and Milwaukee. RAM Broadcasting, Avenel, N.J., is another of the WUTC associates in cellular operations.

WU proposes initial service coverage in Buffalo of 87 percent throughout the total CGSA area allowed since remaining sectors do not justify the added expense at this time. The total is divided into four sets of frequencies, with one set having additional channels over the others because of anticipated higher usage, which is primarily downtown. Less populated areas have fewer cells with larger diameters, and will only be increased when 312 available channels are carrying maximum traffic. Existing Western Union facilities such as towers are to be utilized wherever and whenever possible. Cell site overlaps increase with traffic demands and one is available to carry any excess load as subscriber loads become more numerous. A map showing the Buffalo Area and the proposed cell sites and their overlaps in a 39 dBu contour is depicted in Fig. 2-5. While not as plain as it might be due to photographic reduction, the area and outlay are reasonably evident. As you can see, there is a

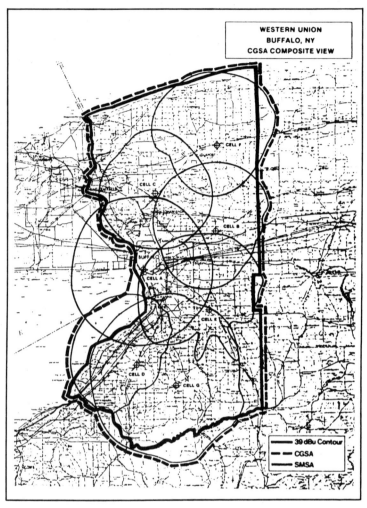

Fig. 2-5. Western Union cellular setup in Buffalo. (courtesy Western Union)

slight breach of the SMSA contours due only to the circular Carey contours which are designed to cover the general area and, according to WU, is de minimis. Any Canadian penetration can be limited by directional antennas even though it will reduce signal strength. If and when performance falls before the 95 percent objective, then immediate remedial action is promised by WU.

Each cell has been planned to contain a data base which will include the number of calls, initial connection, time of call, and call length. All of which will be automatically recorded by a Mobile

Telephone Exchange and processed to determine grades of service, traffic patterns and distribution.

Any required expansion plan calls for additional channels to be added to cells or borrowed from other cells, and even sectorized to a smaller segment of cells. Average call duration has been projected as 94 seconds, with actual connection time of 120 seconds. To meet such requirements for the area, 14 channels were assigned to the Buffalo cell site, 11 to the Clarence cell site, 9 to the Wheatfield cell site, 5 to the Eden, 6 to Elma, 7 to Lockport, and 5 to New Oregon. WU projects that frequency reuse will not be required until the 9th year of operation when all frequency assignments will have been utilized.

Systems surveillance is to come from the primary control point in Buffalo, staffed on a 24-hour basis and equipped with state-of-the-art electrical alarms and test facilities for the entire system. WU claims that problems arising may be quickly identified and repair procedures begun promptly. There is also a secondary alarm center at McGraw, New York which will also have the same alarms and test facilities as Buffalo and will be manned night and day ensuring maximum service reliability and availability. Initial construction cost of the Buffalo system is estimated at $4.534 million.

System Design and Configuration

WU's proposal includes six directional receive antennas in each cell offering 360° coverage and a single transmit antenna with omnidirectional coverage (Fig. 2-6). Receive antennas are rated at 7 dB more gain than the transmitting antenna, permitting mobile and portable subscriber transceivers to operate at lower power. The system is controlled by a Mobile Telephone Exchange that operates at a Class 5 level (or higher) in the Public Telephone Switched Network. The grade of service anticipated should be better than 95 percent and compare directly with that of the public switched telephone network, according to Western Union.

Channels are trunked within cells and system and all are assigned on a demand basis for the various subscribers, including adjacent cell channels during times of overflow. Initial cell radius measures 9 miles with the orderly addition of new cells at required locations already anticipated. The objective being to offer fully coordinated nationwide service for subscribers that is completely automated to the maximum extent possible.

Projected service for Buffalo CGSA (Cellular Geographic Service Area) amounts to 172 voice channels, and since voice channels

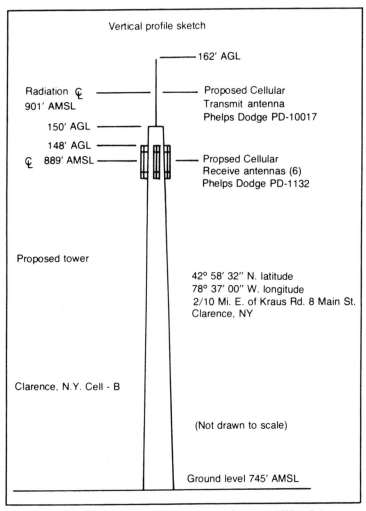

Fig. 2-6. Receive and transmit antennas for WU's Clarence, N.Y. cell. (courtesy Western Union)

do not exceed the 312 nominally available in band A, frequency reuse is not immediately required. Additional channels will be added to each cell site when necessary. This channel addition is illustrated in Fig. 2-7 which may be expanded to more than 100,000 subscribers if needed. Expansion plans for more than 100,000 subscribers also include channel additions, frequency borrowing, more cell sites, sector transmissions, and cell splitting. WU points out that the distance between cells (D) using the same frequency and cell radius (R), amounts to a D/R ratio of between 35 and 60 miles, permitting reuse of certain channels within the serving area. Outside these limits, the subscriber is handed off by cell switching equipment that allows continuous coverage. Directional antennas in sector arrangements aid frequency reuse as applied to the practice of channel borrowing. This permits overall system operation that can both increase and decrease channel capacity in various areas

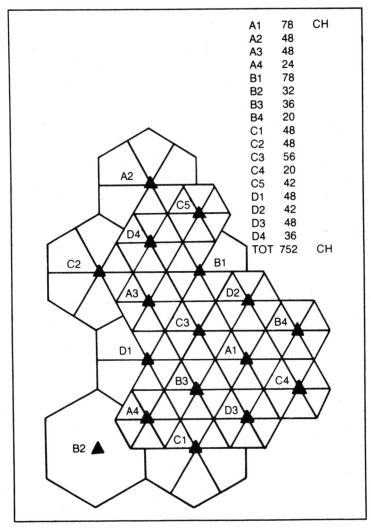

Fig. 2-7. Western Union's second level channel additions for Buffalo. (courtesy Western Union)

where required rather than on a rigid, evenly distributed cellular basis, thereby serving varying numbers of subscribers wherever located.

Cell splitting will take place when system congestion becomes apparent and interferes with the designed grade of service. A horizontal radiation pattern for a Phelps Dodge 1110 cell site transmitting antenna is shown in Fig. 2-8. As you can see, its polarization is horizontal since the omni-pattern is perfectly circular. Note that results are given in 0.5 dB increments.

When WU designs a cellular system, a theoretical grid of cell site locations is used to identify and locate ideal cell sites for best spacing between co-channel cells for frequency reuse. This permits evolution of additional cells and/or smaller cells connected with

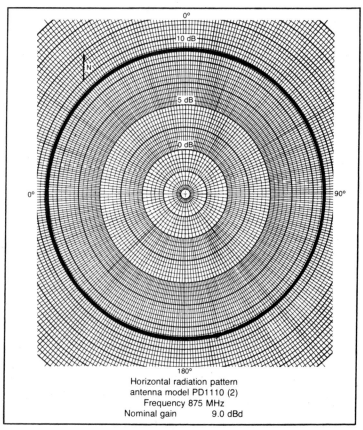

Fig. 2-8. A Phelps Dodge 1110 radiation pattern display. (courtesy Western Union)

frequency reuse. In the interim, any cell may have its capacity increased by borrowing from another for the time being. Later, such borrowed channels may be returned to their original cell and the affected cell may be subdivided and expanded accordingly. Old cells, of course, will be overlapped by new cells, increasing area capacity and also backup capability in the event of subsystem failure. Here, sectored transmissions with directional antennas appear so that reused frequencies will not interfere with one another.

Once area and distribution information has been determined, cells are located and channel plans are formulated using Erlang B equations for cell capacity. Channels are then assigned such as A1, A1/B1, A1/B2, etc., for highest density and highest user subscribers. Should new cells be required, they would look like those in Fig. 2-9, where two new cells are added at high density locations in the pattern. Here, the various cells are omnidirectional. Western Union says that by adding new cells as an overlap, capacity is increased some 20 percent. Should the area continue to grow, frequency reuse will have to be enlisted to further increase capacity of the system.

Final expansion calls for the introduction of additional cell sites, with subdivisions in almost all categories. Limitations here will depend on terrain, area buildings, subscriber numbers and overall operations. After this, only an extension of CGSA can afford additional coverage.

Carey Curves

Western Union also has an interesting application of Carey Curves which may be of profit to some using the 100W ERP calculated versus the 1000 W Carey Curve at 39 dBu and 49 dBu, all in miles. A copy of the calculations follows:

$$D = 10^{**} ((PEFF-38.81+18.77(\log(EAH)))/53.84)$$

where D = Distance to the 39 dBu contour in miles.
** = Exponent of 10.
PEFF = Effective radiated power in dBm.
EAH = Effective antenna height over terrain averaged between 2 and 10 miles from base site in feet.
log = Logarithm to the base 10.

A comparison of values obtained directly from the Carey curves with those calculated using the previous formula is given in Table 2-1.

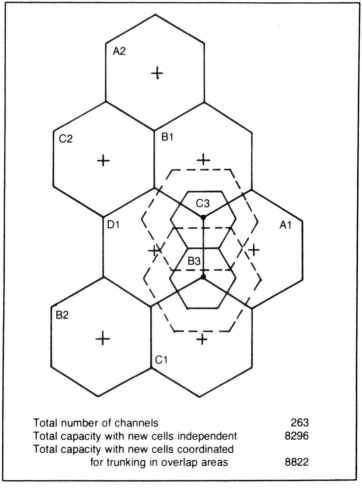

Fig. 2-9. Cell layout with two cells added at high density points in the pattern. All cells are omnidirectional. (courtesy Western Union)

For regular terrain, the relationship between effective height and distance predicted to the 39 dBu contour for the Carey Curves is normal. But where there is irregular terrain, WU says it's possible "to get negative values of effective antenna heights in a radial direction." It may help to extend both a least squares fit and the distance to the nearest line of sight obstruction, but WU claims it's more effective to default to a HAAT of 0.1 of the antenna height above site terrain level. The discontinuity effect is minimized, they say, and contour distance "is easily traced to site data."

Traffic Analysis

To avoid unnecessary expansion, especially in the beginning, WU will monitor and record pattern usage and distribution data as follows:

- ☐ Number of completed calls.
- ☐ Connection time.
- ☐ Call duration.
- ☐ Handoff time duration.
- ☐ Cell receptions and transmissions.
- ☐ Subscriber location.

Western Union will then average the monthly flow of busy hour traffic and determine if each cell meets the 95 percent service objective. If not, WU will immediately expand its coverage to bring any overloaded cell up to standard by taking the following actions:

- ☐ Add nonused channels.
- ☐ Add channels by borrowing from another cell.
- ☐ Implement transmit sectorizing to supply more channels.
- ☐ Cell splitting.

These actions, WU says, will result in high quality and low cost service to its subscribers.

Assignment of channels to various cell sites in the Buffalo area is based on a 5-year forecast and is illustrated in Fig. 2-10. Channel assignments, according to WU, are specifically chosen to permit cell subdivision within a 4-cell reuse frequency pattern. Primary control for the entire system originates at 14 Lafayette Square, Buffalo. Cell sites are equipped with automatic alarms which will notify operators at the control point if any base station transmitter is malfunctioning or exceeding authorized parameters. Such transmitters may then be shut down until the trouble is cleared in one

Table 2-1. Carey Curve Values Versus Calculated Values.

Mean antenna height (feet)	D (Calculated) 100 W ERP 39 dBu (miles)	D (Carey Curve) 1000 W ERP 49 dBu (miles)	Error
100	8.04	8.3	−3.1%
200	10.23	10.4	−1.6
500	14.09	14.0	+0.6
1000	17.94	18.1	−0.8
2000	22.85	22.5	+1.5
5000	31.44	31.0	+1.4

		NUMBER OF CHANNELS (Includes Signaling Channels)				
CELL		YEAR 1	YEAR 2	YEAR 3	YEAR 4	YEAR 5
BUFFALO	- CELL A	15	19	25	34	47
CLARENCE	- CELL B	12	15	19	26	35
WHEATFIELD	- CELL C	10	12	16	21	28
EDEN	- CELL D	6	8	9	12	15
ELMA	- CELL E	7	9	11	14	18
LOCKPORT	- CELL F	8	10	12	16	21
NEW OREGON	- CELL G	6	8	9	12	15
	TOTAL	64	81	101	135	179

Fig. 2-10. Channel assignments for entire Buffalo area. (courtesy Western Union)

way or another. Operator duties include repairs of failed units, spare parts inventory, updating data bases, and software development and applications. Control and base stations are interconnected via transmission links so that major and minor alarms may be continuously monitored when and if they occur. A single voice channel loss, for instance, would trigger a minor alarm, and a fire would be a major alarm. Self diagnostic aids are also to be located at the individual cell sites to reduce or eliminate evident or potential problems before they affect subscriber service.

The Mobile Telephone Exchange will look for the following and register an alarm where there are faults:

☐ Data patterns will check all memory cells.

☐ Software tasks for time completion.

☐ Tone verification, timing sequences, and frequency are monitored.

☐ Redundant equipment automatically goes on line if the data link fails.

☐ Processing of simulated calls for system checks and completions.

☐ Commercial ac power failure should result in generator and battery backup.

☐ Internal supply voltages also monitored.

☐ Any circuit path breakdowns should cause an alarm.

In addition, base station alarms will reach the control point when:

☐ There are failures in frequency sources or they are unstable. Drifts of more than 0.5 ppm will set off an alarm.
☐ Failure of Supervisory Audio Tone.
☐ Excessive temperature in the power amplifier.
☐ Rf power levels generate alarms if the decrease is as much as 25 percent.
☐ Local oscillator or receiver mixers fail.
☐ Battery voltage drops 10 percent.
☐ Commercial power failure.
☐ Base site controller breakdown.
☐ Scan receiver, controller fault.
☐ Signaling channel or its controller presents problems.
☐ Voice channel failure.
☐ MTE to cell site data link interrupts.

Western Union also has an alarm center located at its Manned Microwave Junction Control Station 1.5 miles SSE of McGraw, New York. Continuously staffed, all alarms and controls at the primary control point will also be monitored here, affording an additional level of system surveillance and backup. Maintenance and transiting personnel may be contacted via mobile radiotelephone from this station which also is a maintenance repair and parts depot. The National Facilities Management Center of Western Union headquarters at Upper Saddle River, New Jersey, will also have a record of all system outages and problems, and a daily report of system performance is circulated to senior management.

NASA

Believing that "terrestrial mobile services will not be economically feasible in vast areas of the nation containing 60 million persons," due to rural environment, the National Aeronautics and Space Administration on Nov. 29, 1982 applied to the FCC for "early establishment" of a commercial Land Mobile Satellite Service. In addition to consumer urban needs, NASA also cited other organizations such as national security, law enforcement, emergency medical, interstate transportation, power, and petroleum as being widely-dispersed entities that could use cellular satellite facilities.

FCC Relents

Initially, the FCC flatly refused NASA's suggested shift of the upper part of the cellular band from 890 to 896 MHz on grounds of extended delay because of "time-consuming international negotiations." Canada, it noted, has recently issued a discussion paper on cellular communications suggesting a potential "but localized" conflict among fixed and cellular spectrum assignments between 890-896 MHz that could not be resolved except by frequency reassignments. Mexico also has no agreement with the U.S. in the 890-896 sector, and so cellular could not be used near either border in the near future if the requested frequency shift was approved. The FCC further noted that Canada is now preparing for an 8 MHz mobile satellite system itself.

At the same time, the FCC also turned down NASA's proposal that the 20 MHz reserve portion of the spectrum be consolidated into two 10 MHz bands to accommodate a "possible satellite-based extension of terrestrial based cellular and non-cellular mobile services." The Commissioners decided there was no need at this time for such a satellite service. Further, by leaving the reserve bands intact, the FCC said it was not precluding future development of some mobile satellite system. NASA, for its part, argued that the reserve band order actually precludes some 12 MHz of the 20 MHz reserve band from ever being allocated to mobile satellite service. And without some satellite system, NASA argues, there can be no nationwide service. The four reserve bands set aside are: 821-825 MHz, 845-851 MHz, 866-870 MHz, and 890-896 MHz. The 1979 WARC meeting of participating nations allocated the 806-890 MHz band to mobile satellite services, which NASA would like to have included in 821-831 MHz and 866-876 MHz. NASA has also projected a 400-thousand to 900-thousand subscriber use for satellite-based mobile telephones in rural (non-metropolitan) areas.

Subsequently, the FCC relented after NASA supplied additional information and the results of certain studies which were formerly underway. As a result, a request for comments has already gone to industry from the Commissioners and a decision on a yes or no basis may be forthcoming. At the moment, of course, I don't know what the final results may be, but considering FCC trends toward a free marketplace, one might assume that NASA's request could well be honored.

NASA would have frequencies between 821-825 MHz and 866-870 MHz allocated for this purpose. As in the former proposal, however, NASA does not now propose to disturb the 800 MHz reserve frequency allocation, leaving it as is, and thereby reducing the objections of several FCC Commissioners.

In addition, Canada's Department of Communications (DOC)

and NASA have now agreed in principle to a joint Mobile Satellite Experiment (MSAT-X), and frequency clearances have been requested by both DOC and NASA for a 1987 launch. NASA wants 35-80 MHz of spectrum in either S or Ku bands for feeder links between mobile and fixed satellite earth stations which would be interconnected by Public Telephone Switched Network and/or base stations.

NASA further argues that 8 MHz among the 821-870 MHz frequencies requested will not be able to meet demand by 1995 and that the full 20 MHz reserve will be required by then, following NASA-DOC empirical testing. With this in mind, NASA also proposed to the FCC that it consider L-band use in the 1.5-1.6 GHz range for satellite-mobile communications and submitted a study for Commission perusal. A 20 MHz segment of L-band is suggested for the proposed service. In the meantime, NASA continues to study future private land mobile telecommunications requirements by examining more efficient use of spectrum, newer modulation techniques, reduced channel bandwidth, frequency sharing, and further digital technology.

NASA suggests the following frequencies be set aside:

UPLINK			DOWNLINK		
S-Band	2655-2690	MHz	S-Band	2500-2655	MHz
Ku-Band	12.75-13.25	GHz	Ku-Band	10.7-11.7	GHz
	14.4-14.1	GHz		11.7-12.1	GHz

In the planned U.S.-Canadian experiment, DOC is expected to ask for 13.2-12.25 GHz as the feeder unlink and 11.65-11.7 GHz for the downlink. A total of six spot beams—two of which are to cover the continental U.S.—will operate at the stated frequencies of 821-825 MHz and 866-870 MHz, supported by a single feeder link for both Canada and the continental U.S. Five MHz of feeder link spectrum with a frequency reuse factor of 1.25 is projected initially. Later, possibly a 15-to-30 UHF spot beam system supported by only a few feeder link spot beams would come into being followed by an 87 UHF spot beam configuration with 25 feeder link spot beams. This approach, shown in Fig. 2-11, would require as much as 80 MHz of spectrum according to NASA's Jet Propulsion Laboratory.

System Design and Configuration

This information comes largely from CalTech's Jet Propulsion Laboratory in Pasadena, CA. The study, conducted for NASA dur-

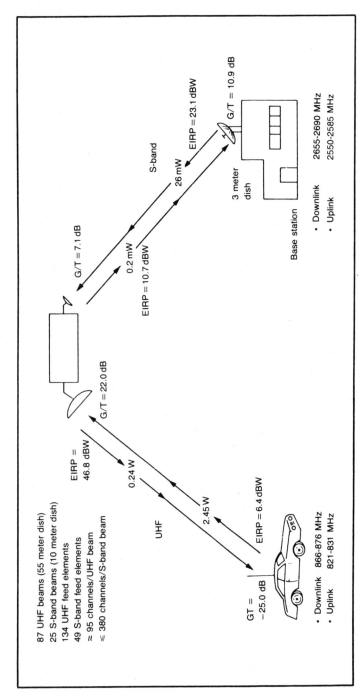

Fig. 2-11. The main NASA satellite-mobile earth station scheduled for 1995 as originally conceived. (courtesy NASA)

ing 1981-82, was undertaken to hypothesize functional requirements for the Land Mobile Satellite Service (LMSS) in providing such telecommunications as paging, dispatch, data exchange, and telephone service to mobile users. Space and ground segments are included.

The study concentrates on a mobile satellite system consisting of one geostationary satellite positioned at 110° WL (west longitude), having a 55-meter UHF antenna and a 10-meter S-band antenna. The system concept is outlined in Fig. 2-11. The continental U.S. (CONUS) would be covered with 87 UHF spot beams, along with 25 S-band spots, producing 8,265 duplex voice channels serving 270 thousand radio telephone users. A full 30 kHz bandwidth/channel was assumed.

Also assumed is the availability of two 10 MHz bands among the 800 MHz sector, and a pair of 35 MHz bands between 2500 and 2690 MHz for feeder-links. All of this is technically feasible by at least 1990. In the meantime, Canada and NASA are planning the joint experimental option of using the 821-825 MHz and 866-870 MHz bands for initial mobile services. Under this arrangement, the operating satellite would have a 10-meter antenna with six spot beams: two for U.S. CONUS, and four for Canada, with Canada sharing one for Alaska.

A single beam feeder-link operating at 13.20-13.25 GHz for uplink and 11.65-11.70 GHz for downlink is planned in the Ku band. NASA filed applications for this service with IRAC and the FCC on September 9, 1982. Objectives include the following:

☐ Tests and evaluations of mobile terminal and base station technology.
☐ Estimation of market size and distribution.
☐ Link performance tests, in addition to cochannel and adjacent channel interference, multipath fading, modulation, mobile unit speeds, and mobile antennas.
☐ User reaction during network blockage, channel access, voice quality, and interference effects.
☐ Network control, switching, phone company interfaces, channel allocations.
☐ Defining technical characteristics within regulatory and institutional parameters.

At the same time, NASA has a contract with Citibank, N.A. for a three-phase financial analysis of the proposed Mobile Satellite Service under various assumptions, using studies previously prepared by other NASA contractors. There are four postulated systems to be studied with estimated costs in terms of 1980 dollars.

According to NASA, the same loading rate, regardless of markets, would be the same for the Citibank evaluation, and that the satellite would not reach full capacity until its 7th year. Systems 3 and 4 were found acceptable and exceed "the derived acceptance standards for investments with compatible risk/reward characteristics of most of the classes of potential investors."

The LMSS Network. The Land Mobile Satellite Service, as shown in Fig. 2-12, consists of the spacecraft, base and mobile stations and their interface with the wireline network. As usual, there is paging, channel assignments, handoffs, signaling and the like. In this design, JPL has presented to Headquarters a 25-base station arrangement with UHF and S-band receptors, transmitting and receiving messages via the various beams and their S-band transmitters, or UHF signals directly to the satellite. Typical cells often begin from the home telephone to wireline networks and thence to a base station for transmission to the satellite over a backhaul link—in this instance over S-band, although C, Ku or Ka bands may also be used, depending on system design. The satellite translates this backhaul information to UHF and relays the call to its ultimate mobile user. When calls originate with the mobile, information goes uplink to the satellite, is translated in frequency, and returns to earth via S-band and backhaul, or it may go first to a base station and thence to the satellite and a base station. The information is then forwarded to some wireline network and the home telephone. Calls may also be routed to other mobile or fixed phones, depending on the originator's address signal. Although NASA does not specifically state what seems evident, a non-wireline system that's connected to the phone network could also receive and broadcast this information with pretty much the same hookup except more microwave would be used instead of land lines.

In its original proposal, NASA had suggested that frequency allocation for the 806-902 MHz band be changed so that additional spectrum might be used. Figure 2-13 illustrates the NASA proposal. The FCC, however, balked, and so the original and current FCC frequency allocation remains, primarily because of delays occasioned by probable foreign negotiations that might take months and even years to resolve. Originally, NASA wanted base-to-mobile and mobile-to-base station frequencies as follows:

- ☐ Base-to-satellite 2655-2690 MHz.
- ☐ Sat.-to-mobile 866-876 MHz.
- ☐ Mobile-to-satellite 821-831 MHz.

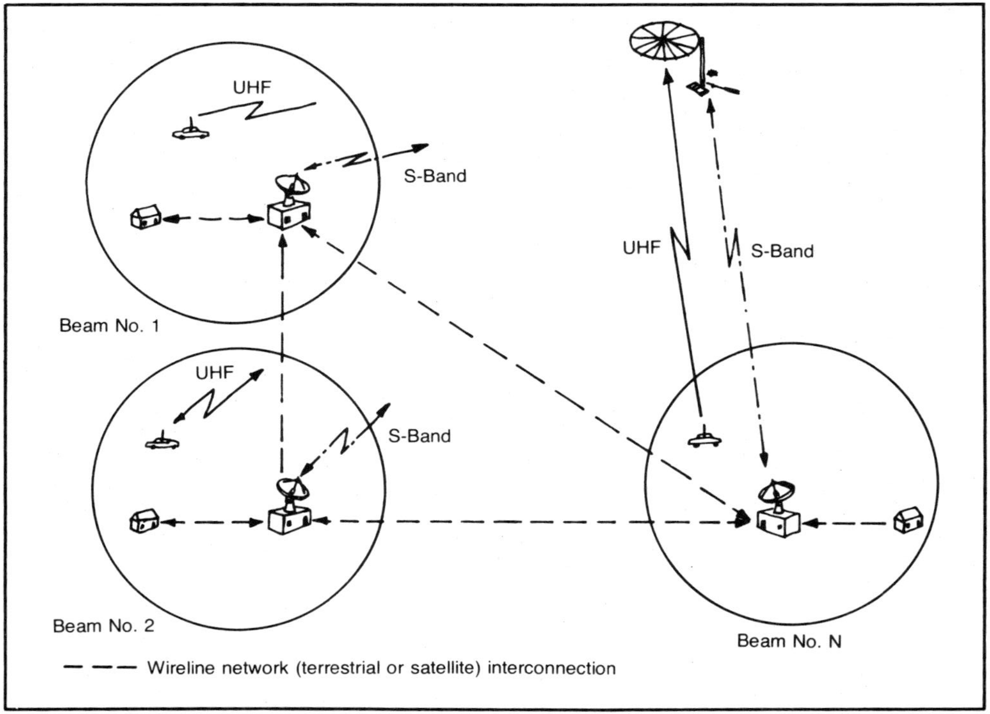

Fig. 2-12. Signal routing in a land mobile satellite system.

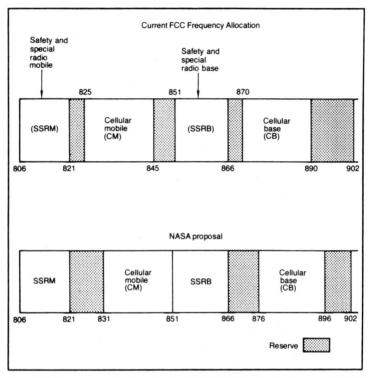

Fig. 2-13. Frequency allocation for the 806-902 MHz band. (courtesy NASA)

☐ Sat.-to-base 2550-2585 MHz.

Uplinks and downlinks, including S-band backhaul are now changed to Ku, especially to accommodate the initial experiment attempt with Canada using either a European L-Sat or some U.S. spacecraft to be determined. Final allocations, of course, will be made by the FCC, taking into account the several requests NASA has already suggested.

Returning to the original four systems proposed for Citibank's initial evaluation, NASA is now offering an independent analysis, assuming 10 MHz of UHF bandwidth, 15 kHz channel spacing, and a 93-beam system having a frequency reuse factor of four. Such a system would allow 15,257 channels and serve 375,000 mobile phone users. At $140 per month for a typical mobile subscriber and $23 per month for a private subscriber, NASA says the system could generate $630 million annually if fully loaded. Using Citibank's 7-year saturation and a still more conservative estimate, of a 65 percent return rate of this figure over system life, the net return

Table 2-2. NASA's Four Postulated Systems Under Study by Citibank.

System	No. Beams	Satellite Antenna Dia.	Power At Startup	Channel Bandwidth	Duplex Voice Ch.	Dollar Cost
1	1	2.4×4 m	5.4 kW	5 kHz	200	$238M
2	13	16.1 m	7.9 kW	15 kHz	670	$375M
3	40	32.5 m	12.8 kW	15 kHz	4.470	$498M
4	100	55 m	16.1 kW	15 kHz	14.000	$678M

would amount to between 25 percent and 29 percent on a gross of $3.8 billion. Satellite design life is an assumed 10 years.

Full duplex radio service has been designed into the system, so any LMSS subscriber can access any telephone connected to the phone network. Further, two mobile users in LMSS may talk directly to one another, but one will have to receive from a base station on the downlink (double hump). Overall, equipment should be quite similar to proposed terrestrial base, hand-held, and mobile stations, and voice channel qualities are expected to compare favorably with normal phone service. An all-digital system, however attractive, will not be instituted at this time because of overall cellular system compatibility.

With respect to ownership, NASA offers the initial COMSAT structure as a possible model, where COMSAT acts as a carrier's carrier with member carriers allowed ownership interests. Terrestrial cellular companies might also procure services from the system for resale to their own subscribers, and also own a part of the system, too. As NASA emphasizes, however, these are only suggestions which the FCC will have to deal.

MSAT. These are the initials of NASA's mobile satellite to be located at 22.3 kilomiles above earth in equatorial orbit and stationed at 110° West longitude. Figure 2-14 and Fig. 2-15 illustrate the MSAT concept.

Expected to weigh some 4,000 kilograms (8.8 K Lbs.) when launched about 1995 via NASA's shuttle, the satellite will have two long booms, one 34 meters vertically and the other 83 meters horizontally to support the 55-meter UHF reflector and the 10-meter (or other) dish.

There are also thermal radiators, as well as a double strip of solar panels supplying 19 kilowatts of source operating power.

The UHF dish is an offset-fed, mesh-deployable parabolic reflector with considerable gain and supplied with signal from a large planar microstrip feed array of 134 elements. Its rectangular envelope measures 6.9 × 11.4 meters for the 87 UHF beams. In this JPL projection, a total of 8,265 duplex voice channels are furnished,

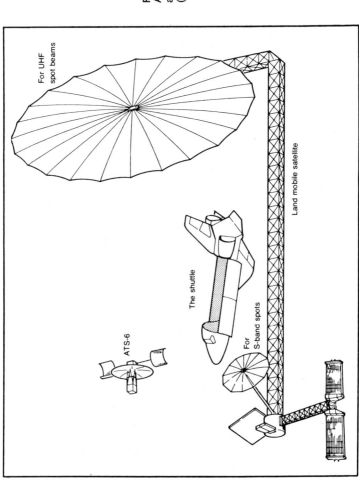

Fig. 2-14. Scaled drawing of MSAT, ATS-6, and the STS shuttle. Note S and UHF spot beam antennas. (courtesy NASA)

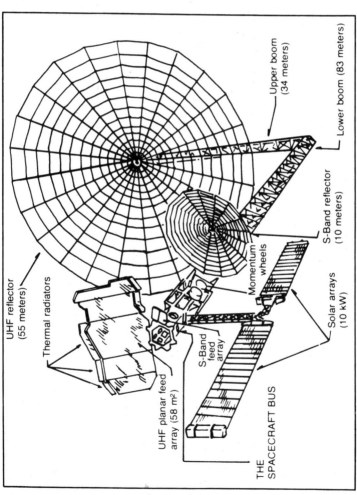

Fig. 2-15. A conceptual drawing depicting MSAT. (courtesy NASA)

each requiring about ¼-watt, or a total rf power need of 2 kW. Operating potential for MSAT's high power amplifiers comes from the two large solar arrays, each of which measures 4 × 9 meters and supplies 10 kW of direct current and voltage.

With its various solar panels, huge reflectors, and mammoth booms, MSAT will be the largest spacecraft ever flown. Consequently, its pointing accuracy (0.03°) and stabilization require state-of-the-art attitude control and special software for in-orbit positioning. Power amplifier heat is disbursed over the 58 meter^2 area located behind the feed array and via heat pipes to remove thermal radiators.

Interbeam isolation for the 87 beams, predicted on a 17 dB net carrier-to-interference (C/I) ratio, has been increased to 27 dB for this design, offering additional margin for reflector surface distortion, misalignments, and uplink interference. This requires patterns with very low sidelobes and multiple feed elements. So the physical diameter is not too great, there is minimum spillover between beams, and also some sharing by adjacent beam feeds. With 7-element clusters, 134 feed elements will produce 87 UHF beams, sufficient to cover at least most of CONUS, at least in theory.

Along with multifeeds and load-sharing, there is also a multiple beam frequency reuse proposal something akin to the cellular system itself. The total frequency allocation for this particular satellite could be divided into a number of sub-bands, each consigned to some designated beam. Then beams with enough separation to prevent signal interaction are also assigned some or all of these same frequencies, permitting the overall band to be reused several times. At the moment, only single polarization of transmitted beams are contemplated because of present microstrip technology, although it is recognized that circular polarization could aid beam isolation, and improve spectrum and communications efficiency.

Antennas and Attitude. The huge 55-meter UHF antenna and its planar feed array uses a wrap-rib technique with ribs, bus assembly, and reflective mesh, all of which can be both unfurled and refurled when and if the occasion demands. A cutaway view is given in Fig. 2-16. Ribs for the offset reflector are constructed of graphite-epoxy for improved thermal and stiffness considerations, while the mesh is specified as 2-bar, tricot knit, gold-plated molybdenum wire. This low stiffness wire permits a 2-directional tension field even with considerable thermal changes and won't develop rf reflective surface wrinkles. It's L-shaped mast is a new technology

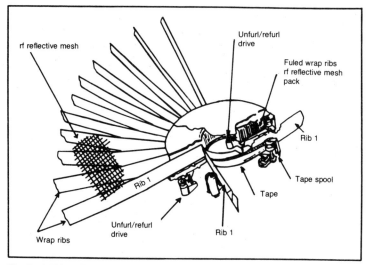

Fig. 2-16. Wrap-Rib reflector concept that may roll up or open. (courtesy NASA)

offering of which we have little information.

Because of its size, very special considerations are required in attitude control design. Advanced software and hardware technologies for system identification, feed motion compensation, and sensing control are all included in the scheme. With a high bandwidth gyro-based control in an attitude determination and gyro drift correction loop using star trackers, primary attitude control is maintained. Stabilization of the reflector and other structural positions are thereafter monitored and controlled by an optical dish sensor located at the feed, with additional control actuators for forces and torques at the feed and hub. There are six of these, which include three fine-pointing reaction wheels and three attitude control and stationkeeping jets at each location. An optical sensor provides alignment checkout and reads dynamic measurements for systems identification.

Three larger wheels permit management of cyclic solar pressure torques and partial management of gravity gradient and balancing torques. Control hardware weighs 1009 lbs, of which 384 lbs. is the propellant needed to combat gravity pull and balance torques. Average power required for these controls is 260 watts.

Up and Downlinks. As Fig. 2-12 shows, there are four links in the Land Mobile Satellite Service, one of which is quite critical and initially can only be reached through a base station. This is the satellite-to-mobile link which, because it is the weakest, becomes the power, thermal, and rf subsystem limitations of the system. Obviously, the mobile antenna is small, vehicles move, receive and transmit paths are omnidirectional, and gain is only 5 dB. Therefore, any satellite attempting to communicate with a mobile has a pretty small target. There is also the problem of fading due to direct path wave attenuation and multipath deflections. With reflected and delayed signals arriving at the same time, the antenna sees delays as well as subtractions and/or additions, depending on reflected energies. However, since this is primarily a rural service, shadowing due to obstructions, is not considered, and the fade margin has been given a design tolerance of 5 dB.

As the diagram shows, each UHF channel requires an average power of 240 milliwatts if using voice operated switching (VOX). In such a system, voice carriers are extinguished when unused, resulting in an overall 4 dB power saving for the satellite. Assuming 50 percent efficient linear amplifiers, MSAT prime power requires some 10 kW for its 8 kilochannels when initially launched.

The User Market

NASA claims that satellite-to-mobile communications are comparable in cost and electronic characteristics to those for similar terrestrial units and that the entire project is quite feasible. It envisages a rural-resident mobile radio-telephone market by 1990 of between 50-288 thousand subscribers, depending on costs and technical qualifications. It also foresees a rural private land mobile radio market in the same year of between 110-976 thousand units based on the same premises and assumptions.

Where the population numbers less than 20 people per square mile, however, NASA does not consider terrestrial systems viable (based on a subscriber rate of 0.5 percent population density and a maximum cell radius of 15 miles). Therefore, areas with less than 20 individuals per square mile would be served only by satellite, and those living outside a small town community of 1 thousand or more would not ordinarily be included in terrestrial cellular systems. NASA estimates that some 40 million people are in this or similar categories versus over 169 million residing within a 566-thousand square mile area that constitutes the 318 Standard Metropolitan Statistical Areas (SMSAs). Hawaii and Alaska are not included in these statistics because of remote design constraints and the great land mass involved. They could, however, still be serviced under special considerations.

The Trucking Industry. Or it may better be identified as the "intercity motor carrier industry," as NASA puts it. And this could well describe just plain, ordinary business people and motorists who have either a desire or need for mobile communications. Unlike existing 2-3 mile Citizens Band Radio (CB), however, this service would be strictly for business instructions and messages, not the totally unlicensed buffalo, beaver, and smokey talk currently occupying channel 19 at 27.185 MHz with AM modulation. Because of cost, only certain trucks would have such expensive $2-$3 thousand mobiles, and their use would, obviously, be quite restricted. Independent truckers possibly would not participate to any extent until mobile prices fell to less than $1 thousand, and that will probably take a while, especially since demand could exceed radio supply for the immediate future.

NASA points out that truckers typically call their home offices often through an 800 number one or more times each day to report route locations and to receive additional instructions. But since motor carriers have now been substantially deregulated, NASA suggests such a procedure is inefficient, especially since truck companies now have authority to divert their carriers for pickups along the way. One national motor carrier estimates that immediate home base-to-driver communications would increase its operating efficiency by as much as 15-20 percent with consequent savings in fuel, repairs, and man hours on the road. As many as 1 million trucks could be affected, and the 318 SMSAs they might use near metropolitan areas would represent only 19 percent of the CONUS land area. So NASA does present quite a cogent argument in support of its extended service area position; and think of all the bus routines and other mobile transport that might benefit also.

One further factor favoring extensive satellite use by trucks might be automatic identification signals from the various trailers so they could be located precisely and almost continuously as they proceed along their designated routes. This might also apply to rolling stock of other carriers, including freight cars and their locomotives. Even aircraft could benefit from such identification and location service. The American Trucking Association reports that some 55 percent of intercity and contract trucks operate in the dispatch mode over long distances.

Other Uses. NASA lists the following additional government and private organizations that could benefit, as well: law enforcement, disaster communications, forest fire reports, emergency medical, search and rescue, and hazardous materials transfer. Although the FCC application does not supply numbers, ATS-3 satellite use in the 1980 eruption of Mount St. Helens was given as a worthwhile example of support for search and rescue team direction and operations.

In a 1979-1980 study, NASA used three VHF channels on ATS-3 along with eight mobile units and four portable units in emergency medical operations in Louisiana, Alabama, and Mississippi. From this experiment, it was determined that mobile services must have interconnects with telephone networks for physician instructions and conferences. There will be an addressable emergency medical market of 64,000 units outside SMSAs by 1990.

The oil and gas industry would also share rather heavily in general usage mobile communications, even more than it does today, and mobile telephones could well become a cost-saving boon to the industry. New service voice and data base units would amount to almost 60-thousand new units by the year 2,000, while new mobile systems among truck tractors and trailers might number over 300 thousand by the turn of the century, according to a contractor's survey for NASA.

Chapter 3

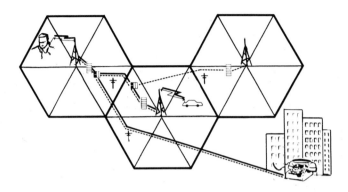

Pertinent Specifications

Information offered in this chapter has been largely derived from the combined EIA (Electronics Industry Association) and FCC Cellular System Mobile/Land Station Compatibility Specification issued in April 1981 and updated with currently available information. Unfortunately, only so much interpolation is allowed, and engineering terms must be followed if the industry is to talk the same language. Consequently, I have tried to make the ensuing specifications readily digestible within the limits of conventional engineering discussion, but have made every attempt to present total fact and no fiction. I do skip around just a little for continuity, and even copy the digital codes outlined by our sources since there is no room for improvement there.

You will find most if not all of the pertinent specifications included, which also embraces the theory of operation for much of the engineering discussion. Therefore, a good deal can be learned from what may be to some rather dry reading. Conversely, those who want the hard facts have them all here in reasonably digestible sequence and continuity. The Signaling Formats with their various codes and logic diagrams appear exactly as they do in the EIA/FCC text since there's little I could do to add to either interpretation or clarity. Sorry to say, these are all rather cut and dried topics where digits and symbols only prevail.

There is, however, a wealth of information dear to both the technical user and the designer among the ensuing pages. I begin, of course, with the land/base transmitter and continue through the mobile transceiver and all the commands and queries it must process in executing the various instructions and responses required in cellular communications. A list of definitions for terms and acronyms appears at the end of the chapter. I trust you will find these specifications all worthwhile.

LIMITATIONS ON EMISSIONS

Wherever you see the above heading in the EIA-FCC cellular guides and regulations, the following will generally or specifically apply, unless you wish to use the FCC published specs verbatim.

For Transmitters (F3) Mobiles - Audio Portion

3 kHz to 15 kHz and 6.1 kHz to 15 kHz, use 40 log F/3.
5.9 kHz to 6.1 kHz 35 dB attenuation is specified.
Above 15 kHz, the attenuation becomes 28 dB.
All this is required after the modulation limiter and before modulation stage.

For Transmitters (F3) Base Stations - Audio Portion

3 kHz to 15 kHz, use 40 log F/3.
Above 15 kHz, attenuation required 28 dB.
Attenuation after modulation limiter; but no notch filter required.

Rf Attenuation Below Carrier Transmitter (F3) - Audio Portion

20 kHz to 40 kHz, use 26 dB.
45 kHz to 2nd harmonic, the specification is 60 dB or 43 + 10 log of mean output power.
12 kHz to 20 kHz, attenuation 117 log F/12.
20 kHz to 2nd harmonic, there is a choice: 100 log F/11 or 60 dB, or 43 log + 10 log of mean output power, whichever is less.

Wideband Data

20 kHz to 45 kHz, use 26 dB.
45 kHz to 90 kHz, use 45 dB.
90 kHz to 2nd harmonic, either 60 dB or 43 + 10 log mean output power.

Supervisory Audio Tones

Same F3 measurements for rf attenuation apply.

Signaling Tone

Same as wideband data but must be 10 kHz ± 1 Hz and produce a nominal frequency deviation of ± 8 kHz.

LAND/BASE STATION TRANSMITTER

With 30 kHz channel spacing and base station operation permitted between 870.030 and 889.980 MHz, the station must maintain a channel carrier frequency within ±1.5 parts/million. Translated into percentage (1.5×10^{-4}), this calculates to 0.00015 percent for any of the assigned base channels. Maximum power permitted amounts to 100 watts ERP.

For *voice signals*, the modulator must be preceded by compressor, preemphasis, deviation limiter and post deviation limiter filter. In the compressor, a 2:1 syllabic compandor which reduces input levels by half, or 2 dB for each 1 dB input. Full response for input signals are to be effected in 3 milliseconds, with full recovery within 13.5 msec. A 1 kHz acoustic tone at the input should produce a ±2.9 kHz frequency deviation of the selected carrier.

Preemphasis should supply a normal +6 dB per octave response between 0.3-3 kHz. For all voice inputs, land stations are required to limit instantaneous frequency deviations to ±12 kHz, excluding supervision and wideband data signals. The deviation limiter requires a following low-pass filter with attenuations that exceed 40 log F/3000 dB in the frequency band between 3 kHz and 15 kHz. Above 15 kHz, the attenuation, relative to 1 kHz, is no less than 28 dB.

Wideband data streams are to be encoded so that NRZ (non-return-to-zero) binary ones and zeroes are now zero-to-one and one-to-zero transitions respectively. Thereafter, wideband data can modulate the transmitter via binary frequency shift keying, and ones and zeroes into the modulator must now be equivalent to nominal peak frequency deviations of 8 kHz above and below the carrier frequency.

Modulation products ±20 kHz from the carrier are to remain at least 26 dB below the unmodulated carrier; those ±45 kHz from the carrier have to be 45 dB or greater below the unmodulated carrier; and others ±90 kHz from the carrier must either rest at 60 dB or more below the unmodulated carrier or 43 plus 10 log mean output power in watts (dB) below the unmodulated carrier.

Spurious conducted and radiated emissions all adhere to current FCC rules, as well as radiations from colocated transmitters with their specific spurious and harmonic levels, all of which are FCC specified.

On voice channels, a supervisory audio tone (SAT) at 5970, 6000 or 6030 is to be modulated on the carrier. Steady state phase error shall be within 0 to ±20°, phase step response is to settle within 10 percent of final steady phase error in at least 250 msec, and the tone modulation index must be within ⅓ radian (±2 kHz) and ±10 percent.

LAND/BASE STATION RECEIVER

Receive transmissions are permitted between 825.030 and 844.980 MHz, with the usual separation for wireline and non-wireline systems via frequency offsets.

Voice signals are to be deemphasized and expanded following the usual FM demodulation. Between 0.3 and 3 kHz, deemphasis occurs at −6 dB/octave; while the syllabic compandor expands received information by 2 dB for every 1 dB input. Expandors are to have attack and recovery responses of 3 and 13.5 msec, respectively. Nominal input reference level to the expandor corresponds to a 1 kHz tone, producing a ±2.9 kHz peak deviation.

Rf signals emitted by the receiver within the mobile station receive band have to be at least 80 dBm down when measured at the antenna connector, and any signals within the mobile transmit band should not exceed −60 dBm at the antenna connector.

Land stations are not required to respond to messages sent by mobiles, and when considerable traffic or overloads occur, mobiles often are allowed to time out following expiration of their normal timing interval of 60 seconds. After that, the mobile is inhibited from transmitting until further commands are issued in any stations that are software-controlled. However, when a mobile is tuned to a voice channel it must monitor any fade timing status. Should this enable, a fade timer is started from reset, and if it counts to five seconds, the mobile has to turn off its transmitter and enter the Serving System Determination Task mode before switching to another status such as Retrieve System Parameters or Paging Channel Selection.

MOBILE TRANSMITTERS AND RECEIVERS

Mobiles are placed by the FCC into three Class I, Class II, and Class III categories, designated High Power, Medium Power, and Low Power; maximum permitted power being 7 watts, or 8.45

Table 3-1. Mobile Station Automatic Attenuation Levels.

MAC Mobile Attenuation Code	Power Classifications		
	I	II	III
000	6	2	−2
001	2	2	−2
010	−2	−2	−2
011	−6	−6	−6
100	−10	−10	−10
101	−14	−14	−14
110	−18	−18	−18
111	−22	−22	−22

dBW. The three categories and their nominal ERP power outputs are as follows:

Class I .... 4W (6 dBW)
Class II .... 1.6 W (2 dBW)
Class III .... 0.6 W (−2 dBW)

Upon land station command, each mobile transmitter must reduce its power in 4 dB steps from 6, 2, and −2 dBW to −22 dBW as illustrated in Table 3-1.

Mobile transmit and receive channels, of course, are exactly the reverse of those in the land station, including the usual 10 MHz offset frequencies assigned that separate wireline and non-wireline operations. It's probably just as well these frequencies are tabulated here as anywhere since this chapter encompasses both base and mobile transceivers. So such frequency assignments are listed in Table 3-2.

Table 3-2. Wireline and Non-Wireline Frequency Pair Assignments.

Non-Wireline (333 freq. pairs, MHz)
Receiver - 825.030, 825.060 to 834.990
Transmitter - 870.030, 870.060 to 870.990
Wireline (333 freq. pairs, MHz)
Receiver - 835.020, 835.050 to 844.980
Transmitter - 880.020, 880.050 to 889.980
Included in these 666 frequency pairs are 21 signaling pairs:
Non-Wireline (MHz)
Receiver - 834.390 through 834.990
Transmitter - 879.390 through 879.994
Wireline (MHz)
Receiver - 835.020 through 835.620
Transmitter - 880.020 through 880.620
Total bandwidth for cellular amounts to 40 MHz, with a 20 MHz reserve. And each channel, of course, measures 30 kHz.

Identification

So that the mobile will recognize a calling identification, a 34-bit binary ID number is given to each transceiver based on its 10-digit telephone number. Every station also has a 32-bit binary serial number that uniquely identifies it in any cellular system. Factory preset, this latter number is isolated from improper contacts or tampering, and any attempt to change the serial number will probably shut down the station.

MIN2. Here's how IDs derive by assigning 10 binary bits to the area code, which we'll call 301. In analog terms, D_1, D_2, and D_3 becomes $100D_1 + 10D_2 + D_3 - 111$, with any 0 digit assuming the value of 10.

$$100(3) + 10 + 1 - 111 = 200$$

And, since the binary sequence advances by doubling,

$$512\ 256\ 128\ 64\ 32\ 16\ 8\ 4\ 2\ 1$$

it isn't difficult to see that 0011001000 becomes the binary number for the sum of $128 + 64 + 8$ which equals 200. Therefore, MIN2 is encoded accordingly.

MIN1. The first part of MIN1 is simply the second three digits of any phone number, say 261. So $D_1 = 2$, $D_2 = 6$, and $D_3 = 1$ and the previous equation gives

$$100(2) + 10(6) + 1 - 111 = 150 \text{ or } 0010010110$$

But the remainder of the number has to be included also, and so the second part of MIN1 is assigned say, 4291, which then becomes 291 since the initial 4 digit is handled separately and is simply 0100. Again, the equation gives

$$100(2) + 10(9) + 1 - 111 = 291$$

which is now read digitally as 0100100011. And working backwards from the last number to the first number

$$MIN1 = 0010\ 0101\ 1001\ 0001\ 0010\ 0011$$

And this, now, becomes our 24-bit essential portion of the 34-bit binary mobile identification number MIN.

Station Class Mark (SCM)

SCM must also be stored in mobile memory to identify both station type and power rating. Each has four alphanumeric identities, all of which are series or precise opposites of one another and are listed in Table 3-3.

Other Memory and Storage Conditions

In addition to SCM, there are other memory requirements suited to different equipments and applications which will be listed briefly for your general information.

SID/NXTREG stands for next registration and system identification and consists of a 15-bit ID and a minimum of four 21-bit next registration signatures, all of which must be retained for longer than 48 hours when power is off *i-f* the mobile station comes equipped for *autonomous registration*. Otherwise, memory has to be reset to zero upon power resumption.

ACCOLC, for access overload class, requires 4-bit number storage to identify overload class field controls. And a 1-bit access (EX) is needed to tell whether extended address words have to be included in all access attempts.

FIRSTCHP means first paging channel, and an 11-bit word identifies the first paging channel number when the mobile is at home.

SID represents a 15-bit designator that identifies a mobile's home system.

LCO, the local control option, requires enable and disable on the mobile. And either System A or B must have internal mobile station identification as the preferred system. When one is selected, the other becomes disabled, and when the Serving System Status is enabled, mobile stations examine signal strengths on the dedicated control channels assigned to System A. If disabled, signal strengths on System B have to be tested.

Supervisory Audio Tone (SAT)

This tone is transmitted by a land station in three frequencies

Table 3-3. SCM Categories.

Power Class		Station Types	
Class I	XX00	Cont. Transmissions	00XX . . . DTX $\neq$ 1
Class II	XX01	Discont. Transmissions	01XX . . . DTX = 1
Class III	XX10	reserved	10XX, 11XX
reserved	XX11	Letters DTX set to 1 permits use of discontinuous transmissions.	

between 5970 and 6030 Hz, must be detected by a mobile, filtered, and then modulated on voice carrier. But SAT is *not* to be transmitted during wideband data but *is* required during signaling tones.

Mobiles are required to decide whether or not SAT is present according to Table 3-4, in the time element of 250 msec.

SAT transmission considerations have already been given. And when SAT is not detected or determined correctly, a fade timing status must be enabled.

Timers and Transmissions are required to perform correctly in any satisfactorily operating system. Timers must run continuously or a breakdown is assumed and mobile stations should halt operations. A protection circuit should also shut down the transmitter when component failures are indicated. Timer expiration periods amount to 60 seconds max.

MOBILE CALL PROCESSING

Although much of this section deals with prime transmissions from the land/base station, the final results are responses of the mobile stations addressed, and so this information is tabulated and explained at this point in order that both transceiver operations may be correlated and understood. Continuing actions, as you will readily see, are relatively inseparable and not easily discussed as individual entities. Therefore we'll try and describe at least a portion of the normal land station transmissions in sequence and their mobile station responses as they occur. Such narration, we hope, will not be confusing.

Calling

Initially, the land station transmits the first part of a SID1 system identification to a mobile monitoring some control channel, followed by the number of paging channels, a serial number request, then mobile registration set to either 0 or 1. When E is set to 1, mobiles transmit both MIN1 and MIN2 during system access,

Table 3-4. Conditions for SAT Detection.

Incoming Signals	SAT Validation	Frequencies
$f < f_1$	Nonvalid	$f_1 = 5955 \pm 5$ Hz
$f_1 \leq f < f_2$	SAT = 5970	$f_2 = 5985 \pm 5$ Hz
$f_2 \leq f < f_3$	SAT = 6000	$f_3 = 6015 \pm 5$ Hz
$f_3 \leq f < f_4$	SAT = 6030	$f_4 = 6046 \pm 5$ Hz
$f_4 \leq f$	Nonvalid	
No SAT	Nonvalid	

another 1 for discontinuous (DTX) transmissions, read control-filler (RCF) message should be set to 1, and access functions (if combined with paging operations) require CPA field setting to 1. Otherwise, CPA goes to 0.

Receiving

As the mobile enters the Scan Dedicated Control Channels Task, it must examine signal strengths of each dedicated control channel assigned to System A if enabled. Disabled, System B control channels are checked.

Then the values assigned in the NAWC (number of additional words coming) system parameter overhead message train will determine for the mobile if all intended information has been received. An EDN field is used as a crosscheck, and control-filler messages are not to be counted as part of the message. Should a correct BCH (Bose-Chaudhuri-Hocquenghem) code be received along with a nonrecognizable overhead message, it must be part of the NAWC count train but the equivalent should not try and execute the instructions.

Under normal circumstances, mobiles are to tune to the strongest dedicated control channel, receive a system parameter transmission and, within 3 seconds, set up the following:

☐ Set SID's 14 most significant bits to SID1 field value.
☐ Set SID's least significant bit to 1, if serving system status enables, or to zero if not.
☐ Set paging channels N to 1 plus the value of N-1 field.
☐ Set paging channel FIRSTCHP as follows:
   If $SID_s = SID_p$, then $FIRSTCHP_s = FIRSTCHP_p$ (an 11-bit paging channel).
   If $SID_s = SID_p$ and serving system is enabled, set $FIRSTCHP_s$ to initial dedicated channel for System B.
   If $SID_s = SID_p$ and serving system is disabled, set $FIRSTCHP_s$ to first dedicated control channel for System B.
☐ Set $LASTCHP_s$ to value of $FIRSTCHP_s + N_s - 1$.

Should the mobile come equipped for autonomous registration, it must

☐ Set registration increment ($REGINCR_s$) to its 450 default value.
☐ Set registration ID status to enabled.

At this point, a mobile must begin the Paging Channel Selection Task. If this cannot be completed on the strongest dedicated control channel, the second strongest dedicated channel may be accessed and the three-second interval commenced again. Incomplete results should result in a serving system status check and an enabled or disabled state reversed, permitting the mobile to begin the Scan Dedicated control Channels Task when channel signal strengths are once more examined.

Overhead Information

Custom local operations for mobiles may be sent and include roaming mobiles whose home systems are group members. A new access channel, that is not among defaults, may be transmitted with a new access field set to the initial access channel. Autonomously registered mobiles may increment their next registered ID by some fixed value, but the global action message must have its REGINCR field adequately set. Also, so that all mobiles will enter the Initialization Task and scan dedicated control channels, a RESCAN global action message must be transmitted.

Mobile stations may be required to read a control-filler message before accessing any system on a reverse control channel. Usually the RCF field is set to 1, otherwise to zero. And should access functions and paging functions be combined, the combined paging/access (CPA) field has to be set to 1. If access functions are not on the same channels as the paging functions, CPA fields are set to zero. Mobiles may also receive 2-word orders, which amount to either Audit or Local Control.

System access for mobiles is sent on a forward control channel in the following manner. Digital Color Code (DCC) identifies the land station. Control Mobile Attenuation Code (CMAC) is included in the control-filler message for mobile power level transmitter adjustment before accessing any system on a reverse control channel. Here, the RCF field is set to 1, but when power levels are not used or required, the CMAC field is to register 000. The WFOM, Wait for Overhead Message field, must also register 0 before the mobile accesses a system on a reverse control channel. When mobiles assigned to one or more of the 16 overload classes are not to access originations on a reverse control channel, an overload control message is carried with the system parameter overhead message, overload class fields are set to zero among the restricted number, and the remainder set to 1. Busy-to-idle status (BIS) access parameters go to zero when mobiles are prevented from checking on the reverse control channel and the message must be added to the overhead. When mobiles can't use the reverse control channel for seizure message attempts or busy signals, access attempt parameters must also be included in the overhead. And when a land station receives a seizure precursor matching its digital color code with 1 or no bit errors, busy idle bits signals on the forward control channel must be set to busy within 1.2 milliseconds from the time of the last bit seizure. Busy-idle bit then must remain busy until a minimum of 30 msec following the final bit of the last word of the message has been received, or a total of 175 msec has elapsed.

Channel Confirmation

Mobiles are to monitor station control messages for orders and respond to both audio and local control orders even though land stations are not required to reply. MIN bits must be matched. Thereafter, the System Access Task is entered with a page response, as above, and an access timer started.

This timer runs as follows:
- ☐ 12 seconds for an origination.
- ☐ 6 seconds for page response.
- ☐ 6 seconds for an order response.
- ☐ 6 seconds for a registration.

The last try code is then set to zero, and the equipment begins the Scan Access Channels Task to find two channels with the strongest signals which it tunes and enters the Retrieve Access Attempts Parameters Task.

Here, both maximum numbers of seizure attempts and busy signals are each set to 10. Thereafter, a read controller-filler bit (RCF) should be checked: if the RCF equals zero, then the mobile reads a control-filler message, sets DCC and WFOM (wait for overhead message train before reverse control channel access) to the proper fields and sets the appropriate power level. Should neither DCC field nor control-filler message be received and access time has expired, the mobile station goes to Serving System Determination Task. But within the allowed access time, the mobile station enters the Alternate Access Channel Task. BIS is then set to 1 and the WFOM bit is checked. If WFOM equals 1, the station enters the Update Overhead Information Task; if WFOM equals 0, a random delay wait is required of 0 to 200 msec, ±1 msec. Then, the station enters the Seize Reverse Control Channel Task.

Service Request

This task requires that the mobile continue to send its message to the land station according to the following instructions:

☐ Word A is required at all times.

☐ Word B has to be sent if last try access LT equals 1 or if E requires MIN1 and/or MIN2, and the ROAM status is disabled, or if the station has been paged with a 2-word control message.

☐ Word C is transmitted with S (serial number) being 1.

☐ Word D required if the access is an origination.

☐ Word E transmitted when the access is an origination and between 9 and 16 digits are dialed. When the mobile has transmitted its complete message, an unmodulated carrier is required for another 25 milliseconds before carrier turnoff. After words A through E have been sent, the next mobile task depends on the type of access.

Order confirmation requires entry into the Serving System Determination Task.

Origination means entry into the Await Message Task.

Page response, is the same as Origination.

Registration requires Await Registration Confirmation, which must be completed within 5 seconds or registration failure follows. The same is true for Await Message since an incomplete task in 5 seconds sends the mobile into the Serving System Determination Task. Origination or Page response requires mobile update of parameters delivered in the message. If R equals 1, the mobile enters the Autonomous Registration Task, otherwise, it goes to the Initial Voice Channel Confirmation Task. Origination access may be either an intercept or reorder, and in these instances, mobiles enter the Serving System Determination Task. The same holds true for a page response access. But if access is an origination and the user terminates his call during this task, the call has to be released on a voice channel and not control channel.

If a mobile station is equipped for Directed Retry and if a new message is received before all four words of the directed retry message, it must go to the Serving System Determination Task. There the last try code (LT) must be set according to the ORDQ (order qualifier) field of the message as follows:

If 000, LT sets to 0
If 0001, LT sets to 1

Thereafter, the mobile clears the list of control channels to be scanned in processing Directed Retry (CCLIST) and looks at each CHANPOS (channel position) field contained in message words three and four. For nonzero CHANPOS field, the mobile calculates a corresponding channel number by adding CHANPOS to FIRSTCHA minus one. Afterwards, the mobile has then to determine if each channel number is within the set designated for cellular systems. A true answer requires adding this/these channel(s) to the CCLIST.

When a Directed Retry message has been answered, access timing must be checked. If expired, the mobile station enters the Serving System Determination Task. If not, the mobile station goes to the Directed Retry Task.

Awaiting Answers

Here, an alert timer is set for 65 seconds (0 to +20 percent). During this period the following events may take place:

☐ Should time expire, the mobile turns its transmitter off and enters the Serving System Determination Task.

☐ An answer requires signaling tone turnoff and Conversation Task entry.

☐ If any of the messages listed hereafter are received within 100 milliseconds, the mobile must compare SCC digits that identify stored and proper SAT frequencies for the station to the PSCC (present SAT color code). If not equivalent, the order is ignored. If correct, then the following actions are taken for each order:

Handoff. Signaling extinguished for 500 msec, signal tone off, transmitter off, power level adjusted, new channel tuned, new SAT, new SCC field, transmitter on, fade timer reset, and signaling tone on. Wait for Answer Task.

Alert. Reset alert timer for 65 seconds and stay in Waiting for Answer Task.

Stop Alert. Extinguish signaling tone and enter Waiting for Order Task.

Release. Signaling tone off, wait 500 msec, then enter Release Task.

Audit. Confirm message to land station, then stay in Waiting for Answer Task.

Maintenance. Reset alert timer for 65 seconds and remain in Waiting for Answer Task.

Change Power. Adjust transmitter to power level required and send confirmation to land station. Remain in Waiting for Answer Task.

Local Control. If local control is enabled and order received, examine LC field and determine action.

Orders other than the above for this type of action are ignored.

Conversation

In this mode, a release-delay timer is set for 500 msec. If Termination is enabled, the mobile sets termination status to disabled and waits 500 milliseconds before entering Release Task. The following actions may then execute:

☐ Upon call termination, the release delay timer has to be checked. If time has expired, the Release Task is entered; if not expired, the mobile must wait until expiration and then enter Release Task.

☐ Upon user requested flash, signaling tone turned on for 400 msec. But should a valid order tone be received during this interval, the flash is immediately terminated and the order processed. The flash, of course, is not then valid.

☐ Upon receipt of the following listed orders and within 100 msec, the mobile must compare SCC with PSCC, and the order is ignored if the two are not equal. But if they are the same, the following can occur:

Handoff. Signaling tone on for 50 msec, then off, transmitter off, power level adjusted, new channel tuned, adjust new SAT, set SCC to SCC field message value, transmitter on, fade timer reset, remain in Conversation Task.

Send Called Address. Upon receipt within 10 seconds of last valid flash, called address sent to land station. Mobile remains in Conversation Task. Otherwise, remain in Conversation Task.

Alert. Turn on signaling tone, wait 500 msec, then enter Waiting for Answer Task.

Release. Check release delay timer. If time expired, mobile enters Release Task; but if timer has not finished, then mobile must wait and then enter Release Task when time has expired.

Audit. Order confirmation sent to land station while remaining in Conversation Task.

Maintenance. Signaling tone on, wait 500 msec, then enter Waiting for Answer Task.

Change Power. Adjust transmitter to power level required by order qualification code and send confirmation to land station. Remain in Conversation Task.

Local Control. If local control is enabled and local control order received, the LC field is to be checked for subsequent action and confirmation.

Orders other than the above for this type of action are ignored.

Release

In the release mode the following steps are required:

☐ Signaling tone sent for 1.8 sec. If flash in transmission when signaling tone begun, it must be continued and timing bridged so that action stops within 1.8 seconds.

☐ Stop signaling tone.

☐ Turn off transmitter.

☐ The mobile station then enters the Serving System Determination Task.

And that concludes interpolations and additions to the Cellular System Mobile/Land Station Compatibility Specification (OST Bulletin No. 53) dated April 1981. To conclude the chapter, Signaling Formats, Message Confirmations, and allied information will be used verbatim from this specification just as it has been printed. You should find it completely understandable given all foregoing information and explanations, in addition to EIA/FCC definitions that are included at chapter's end.

SIGNALING FORMATS

Reverse Control Channel

The reverse control channel (RECC) is a wideband data stream sent from the mobile station to the land station. This data stream must be generated at a 10 kilobit/second ±1 bit/second rate. Figure 3-1 depicts the format of the RECC data stream.

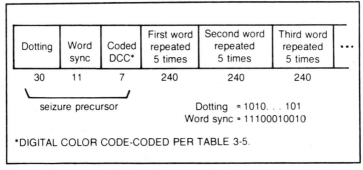

Fig. 3-1. Reverse control channel message stream.

Received DCC	7-Bit Coded DCC
00	0000000
01	0011111
10	1100011
11	1111100

Table 3-5. Coded Digital Color Code.

All messages begin with the RECC seizure precursor which is composed of a 30-bit dotting sequence (1010...101), an 11-bit word sync sequence (11100010010), and the coded digital color code (DCC). The 7-bit coded DCC is obtained by translating the received DCC according to Table 3-5.

Each word contains 48 bits, including parity, and is repeated five times; it is then referred to as a word block. A word is formed by encoding 36 content bits into a (48, 36) BCH code that has a distance of 5, (48 36; 5). The left-most bit (i.e., earliest in time) shall be designated the most-significant bit. The 36 most-significant bits of the 48-bit field shall be the content bits.

The generator polynomial for the code is the same for the (40,28;5) code used on the forward control channel.

Each RECC message can consist of one to five words. The types of messages to be transmitted over the reverse control channel are as follows:

☐ Page Response Message.
☐ Origination Message.
☐ Order Confirmation Message.
☐ Order Message.

These messages are made up of combinations of the following five words:

Word A - Abbreviated Address Word

F = 1	NAWC	T	S	E	RSVD = 0	SCM	$MIN1_{23-0}$	P
1	3	1	1	1	1	4	24	12

Word B - Extended Address Word

F = 0	NAWC	LOCAL	ORDQ	ORDER	LT	RSVD = 00...0	$MIN2_{33-24}$	P
1	3	5	3	5	1	8	10	12

Word C - Serial Number Word

F = 0	NAWC	SERIAL	P
1	3	32	12

Word D - First Word of the Called-Address

F = 0	NAWC	1st DIGIT	2nd DIGIT	...	...	...	...	7th DIGIT	8th DIGIT	P
1	3	4	4	4	4	4	4	4	4	12

Word E - Second Word of the Called-Address

F = 0	NAWC = 000	9th DIGIT	10th DIGIT	...	...	...	...	15th DIGIT	16th DIGIT	P
1	3	4	4	4	4	4	4	4	4	12

The interpretation of the data fields is as follows:

F—First word indication field. Set to "1" in first word and "0" in subsequent words.

NAWC—Number of additional words coming field.

T—T field. Set to "1" to identify the message as an origination or an order; set to "0" to identify the message as an order response or page response.

S—Send serial number field. If the serial number word is sent, set to "1"; if the serial number word is not sent, set to "0."

E—Extended address field. If the extended address word is sent, set to "1;" if the extended address word is sent, set to "0."

SCM—The station class mark field.

ORDER—Order field. Identifies the order type.

ORDQ—Order qualifier field. Qualifies the order confirmation to a specific action.

LOCAL—Local control field. This field is specific to each system. The ORDER field must be set to locate control for this field to be interpreted.

LT—Last-try code field.

MIN1—First part of the mobile identification number field.

MIN2—Second part of the mobile identification number field.

SERIAL—Serial number field. Identifies the serial number of the mobile station.

Table 3-6. Digit Code.

Digit	Code	Digit	Code
1	0001	7	0111
2	0010	8	1000
3	0011	9	1001
4	0100	0	1010
5	0101	*	1011
6	0110	#	1100
		Null	0000

NOTE:
1) The digit 0 is encoded as binary ten; not binary zero.
2) The code 0000 is the null code, indicating no digit present.
3) All other four-bit sequences are reserved, and must not be transmitted

DIGIT—Digit field (see Table 3-6).

RSVD—Reserved for future use; all bits must be set as indicated.

P—Parity field.

Examples of encoding called-address information into the called-address words follow:

If the number 2# is entered, the word is as follows:

*NOTE	0010	1100	0000	0000	0000	0000	0000	0000	P
4	4	4	4	4	4	4	4	4	12

If the number 13792640 is entered, the word is as follows:

*NOTE	0001	0011	0111	1001	0010	0110	0100	1010	P
4	4	4	4	4	4	4	4	4	12

If the number *24273258 is entered, the words are as follows:

Word D - First Word of the Called-Address

*NOTE	1011	0010	0100	0010	0111	0011	0010	0101	P
4	4	4	4	4	4	4	4	4	12

Word E - Second Word of the Called-Address

*NOTE	1000	0000	0000	0000	0000	0000	0000	0000	P
4	4	4	4	4	4	4	4	4	12

*NOTE: These four bits depend on the type of message.

Reverse Voice Channel

The reverse voice channel (RVC) is a wideband data stream sent from the mobile station to the land station. This data stream must be generated at a 10 kilobit/second ±1 bit/second rate. Figure 3-2 depicts the format of the RVC data stream.

A 37-bit dotting sequence (1010 ... 101) and an 11-bit word sync sequence (11100010010) are sent to permit land stations to achieve synchronization with the incoming data, except at the first repeat of word 1 of the message where a 101-bit dotting sequence is used. Each word contains 48 bits, including parity, and is repeated five times together with the 37-bit dotting and 11-bit word sync sequences; it is then referred to as a word block. For a multi-word message, the second word block is formed the same as the first word block including the 37-bit dotting and 11-bit word sync sequences. A word is formed by encoding the 36 content bits into a (48, 36) BCH code that has a distance of 5, (48, 36; 5). The left-most bit (i.e., earliest in time) shall be designated the most-significant bit. The 36 most-significant bits of the 48-bit field shall be the content bits. The generator polynomial for the code is the same as for the (40, 28; 5) code used on the forward control channel.

Each RVC message can consist of one or two words. The types

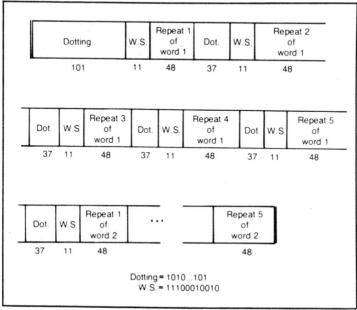

Fig. 3-2. Reverse voice channel message stream.

of messages to be transmitted over the reverse voice channel are as follows:

- ☐ Order Confirmation Message.
- ☐ Called-Address Message.

The message formats are as follows:

Order Confirmation Message

F = 1	NAWC = 00	T = 1	LOCAL	ORDQ	ORDER	RSVD = 000...0	P
1	2	1	5	3	5	19	12

Called-Address Message

Word 1 - First Word of the Called-Address

F = 1	NAWC = 01	T = 0	1st DIGIT	2nd DIGIT	....	7th DIGIT	8th DIGIT	P
1	2	1	4	4	4 4 4 4 4	4	4	12

Word 2 - Second Word of the Called-Address.

F = 0	NAWC = 00	T = 0	9th DIGIT	10th DIGIT	....	15th DIGIT	16th DIGIT	P
1	2	1	4	4	4 4 4 4 4	4	4	12

The interpretation of the data fields is as follows:

F—First word indication field. Set to "1" in first word and "0" in second word.

NAWC—Number of additional words coming field.

T—T field. Set to "1" to identify the message as an order confirmation. Set to "0" to identify the message as a called-address.

DIGIT—Digit field (see Table 3-6).

ORDER—Order field. Identifies the order type.

ORDQ—Order qualifier field. Qualifies the order confirmation to a specific action.

LOCAL—Local control field. This field is specific to each system. The ORDER field must be set to local control for this field to be interpreted.

RSVD—Reserved for future use; all bits must be set as indicated.

P—Parity field.

Forward Control Channel

The forward control channel (FOCC) is a continuous wideband data stream sent from the land station to the mobile station. This data stream must be generated at a 10 kilobit/second ±0.1 bit/second rate. Figure 3-3 depicts the format of the FOCC data stream.

Each forward control channel consists of three discrete information streams, called stream A, stream B, and busy-idle stream, that are time-multiplexed together. Messages to mobile stations with the least significant bit of their mobile identification number equal to "0" are sent on stream A, and those with the least-significant bit of their mobile identification number equal to "1" are sent on stream B.

The busy-idle stream contains busy-idle bits, which are used to indicate the current status of the reverse control channel. The reverse control channel is busy if the busy-idle bit is equal to "0" and idle if the busy-idle bit is equal to "1." A busy-idle bit is located at the beginning of each dotting sequence, at the beginning of each

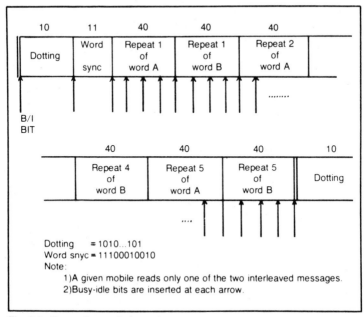

Fig. 3-3. Forward control channel message stream.

word sync sequence, at the beginning of the first repeat of word A, and after every 10 message bits thereafter.

A 10-bit dotting sequence (1010101010) and an 11-bit word sync sequence (11100010010) are sent to permit mobile stations to achieve synchronization with the incoming data. Each word contains 40 bits, including parity, and is repeated five times, it is then referred to as a word block. For a multiword message, the second word block and subsequent word blocks are formed the same as the first word block including the 10-bit dotting and 11-bit word sync sequences. A word is formed by encoding 28 content bits into a (40, 28; 5) BCH code. The left-most bit (i.e., earliest in time) shall be designated the most-significant bit. The generator polynominal for the (40, 28; 5) BCH code is

$$g_B(X) = X^{12} + X^{10} + X^8 + X^5 + X^4 + X^3 + X^0.$$

The code, a shortened version of the primitive (63, 51; 5) BCH code, is a systematic linear block code with the leading bit as the most significant information bit and the least-significant bit as the last parity-check bit.

Each FOCC message can consist of one or more words. The types of messages to be transmitted over the forward control channel are:

☐ Mobile station control message.
☐ Overhead message.
☐ Control-filler message.

Control-filler messages may be inserted between messages and between word blocks of a multiword message.

The following sections contain descriptions of the message formats that the land station transmits over either stream A or B. For purposes of format presentation and explanation, the busy-idle bits have been deleted in the discussion of the message formats.

Mobile Station Control Message. The mobile station control message can consist of one, two, or four words.

Word 1 - Abbreviated Address Word

T_1 t_2	DCC	$MIN1_{23-0}$	P
2	2	24	12

Word 2 - Extended Address Word

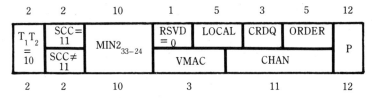

Word 3 - First Directed-Retry Word

T_1T_2 $=$ 10	SCC $=$ 11	CHANPOS	CHANPOS	CHANPOS	RSVD $=$ 000	P
2	2	7	7	7	3	12

Word 4 - Second Directed-Retry Word

T_1T_2 $=$ 10	SCC $=$ 11	CHANPOS	CHANPOS	CHANPOS	RSVD $=$ 000	P
2	2	7	7	7	3	12

The interpretation of the data fields is as follows:

$T_1 T_2$ —Type field. If only Word 1 is sent, set to "00" in Word 1. If a multiple-word message is sent, set to "01" in Word 1 and set to "10" in each additional word.

DCC—Digital color code field.

MIN1—First part of the mobile identification number field.

MIN2—Second part of the mobile identification number field.

SCC—SAT color code (see Table 3-7).

ORDER—Order field. Identifies the order type (see Table 3-8).

ORDQ—Order qualifier field. Qualifies the order to a specific action (see Table 3-8).

LOCAL—Local control field. This field is specific to each system. The ORDER field must be set to local control (see Table 3-8) for this field to be interpreted.

VMAC—Voice mobile attenuation code field. Indicates the mobile station power level associated with the designated voice channel.

CHAN—Channel number field. Indicates the designated voice channel.

CHANPOS—Channel position field. Indicates the position of a

Table 3-7. SAT Color Code.

Bit Pattern	SAT Frequency
00	5970 Hz
01	6000 Hz
10	6030 Hz
11	(Not a channel designation)

control channel relative to the first access channel (FIRSTCHA).

RSVD—Reserved for future use, all bits must be set as indicated.

P—Parity field.

Overhead Message. A three-bit OHD field is used to identify the overhead message types. Overhead message type codes are listed in Table 3-9, and are grouped into the following functional classes:

- ☐ System parameter overhead message.
- ☐ Global action overhead message.
- ☐ Registration identification message.
- ☐ Control-filler message.

Overhead messages are sent in a group called an overhead

Table 3-8. Order and Order Qualification Codes.

Order Code	Order Qualification Code	Function
00000	000	page (or origination)
00001	000	alert
00011	000	release
00100	000	reorder
00110	000	stop alert
00111	000	audit
01000	000	send called-address
01001	000	intercept
01010	000	maintenance
01011	000	charge power to power level 0
01011	001	change power to power level 1
01011	010	change power to power level 2
01011	011	change power to power level 3
01011	100	change power to power level 4
01011	101	change power to power level 5
01011	110	change power to power level 6
01011	111	change power to power level 7
01100	000	directed retry - not last try
01100	001	directed retry - last try
01101	000	non-autonomous registration - do not make whereabouts known
01101	001	non-autonomous registration - make whereabouts known
01101	010	autonomous registration - do not make whereabouts known
01101	011	autonomous registration - make whereabouts known
11110	000	local control

(All other codes are reserved)

Table 3-9. Overhead Message Types.

Code	Order
000	registration ID
001	control-filler
010	reserved
011	reserved
100	global action
101	reserved
110	word 1 of system parameter message
111	word 2 of system parameter message

message train. The first message of the train must be the system parameter overhead message. The desired global action messages and/or a registration ID message must be appended to the end of the system parameter overhead message. The total number of words in an overhead message train is one more than the value of the NAWC field contained in the first word of the system parameter overhead message. The last word in the overhead message train is identified by a "1" in the END field of that word; the END field of all other words in the train must be set to "0." For NAWC-counting purposes, inserted control-filler messages must not be counted as part of the overhead message train.

The system parameter overhead message must be sent every 0.8 ±0.3 seconds on each of the following control channels:

☐ Combined paging-access forward control channel.

☐ Separate paging forward control channel (i.e., CPA = 0).

☐ Separated access forward control channel (i.e., CPA = 0) when the control-filler message is sent with the WFOM bit set to "1."

The global action messages and the registration identification message are sent on an as needed basis.

☐ The system parameter overhead message consists of two words.

Word 1

T_1T_2 = 11	DCC	SID1	RSVD = 000	NAWC	OHD = 110	P
2	2	14	3	4	3	12

Word 2

T_1T_2 = 11	DCC	S	E	REGH	REGR	DTX	RSVD = 0
2	2	1	1	1	1	1	1

N−1	RCF	CPA	CMAX−1	END	OHD = 111	P
5	1	1	7	1	3	12

The interpretation of the data fields is as follows:

T_1T_2—Type field. Set to "11" indicating an overhead word.

OHD—Overhead message type field. The OHD field of Word 1 is set to "110" indicating the first word of the system parameter overhead message. The OHD field of Word 2 is set to "111" indicating the second word of the system parameter overhead message.

DCC—Digital color code field.

SID1—First part of the system identification field.

NAWC—Number of additional words coming field. In Word 1 this field is set to one fewer than the total number of words in the overhead message train.

S—Serial number field.

E—Extended address field.

REGH—Registration field for home stations.

REGR—Registration field for roaming stations.

DTX—Discontinuous transmission field.

N−1—N is the number of paging channels in the system.

RCF—Read-control-filler field.

CPA—Combined paging/access field.

CMAX−1—CMAX is the number of access channels in the system.

END—End indication field. Set to "1" to indicate the last word of the overhead message train; set to "0" if not last word.

RSVD—Reserved for future use, all bits must be set as indicated.

P—Parity field.

☐ Each global action overhead message consists of one word. Any number of global action messages can be appended to a system parameter overhead message.

The formats for the global action commands are as follows:

Rescan Global Action Message

T_1T_2 =11	DCC	ACT = 0001	RSVD= 000...0	END	OHD = 100	P
2	2	4	16	1	3	12

Registration Increment Global Action Message

T_1T_2 =11	DCC	ACT = 0010	REGINCR	RSVD = 0000	END	OHD = 100	P
2	2	4	12	4	1	3	12

New Access Channel Set Global Action Message

T_1T_2 =11	DCC	ACT = 0110	NEWACC	RSVD = 00000	END	OHD = 100	P
2	2	4	11	5	1	3	12

Overload Control Global Action Message

T_1T_2 =11	DCC	ACT = 1000	OLC0	OLC1	OLC2	OLC3	OLC4	OLC5	OLC6	OLC7
2	2	4	1	1	1	1	1	1	1	1

OLC8	OLC9	OLC10	OLC11	OLC12	OLC13	OLC14	OLC15	END	OHD = 100	P
1	1	1	1	1	1	1	1	3	12	

Access Type Parameters Global Action Message

T_1T_2 =11	DCC	ACT = 1001	BIS	RSVD= 000...0	END	OHD = 100	P
2	2	4	1	15	1	3	12

Access Attempt Parameters Global Action Message

T_1T_2 =11	DCC	ACT = 1010	MAXBUSY −PGR	MAXSZTR −PGR	
2	2	4	4	4	

MAXBUSY –OTHER	MAXSZTR –OTHER	END	OHD = 100	P
4	4	1	3	12

Local Control 1 Message

T_1T_2 = 11	DCC	ACT = 1110	LOCAL CONTROL	END	OHD = 100	P
2	2	4	16	1	3	12

Local Control 2 Message

T_1T_2 = 11	DCC	ACT = 1111	LOCAL CONTROL	END	OHD = 100	P
2	2	4	16	1	3	12

The interpretation of the data field is as follows:

$T_1 T_2$—Type field. Set to "11" indicating overhead word.
ACT—Global action field. See Table 3-10.
BIS—Busy-idle status field.
DCC—Digital color code field.
OHD—Overhead message type field. Set to "100" indicating the global action message.
REGINCR—Registration increment field.
NEWACC—News access channel starting point field.

Table 3-10. Global Action Message Types.

Action Code	Type
0000	reserved
0001	rescan paging channels
0010	registration increment
0011	reserved
0100	reserved
0101	reserved
0110	new access channel set
0111	reserved
1000	overload control
1001	access type parameters
1010	access attempt parameters
1011	reserved
1100	reserved
1101	reserved
1110	local control 1
1111	local control 2

MAXBUSY-PGR—Maximum busy occurrences field (page response).

MAXBUSY-OTHER—Maximum busy occurrences field (other accesses).

MAXSZTR-PGR—Maximum seizure tries field (page response).

MAXSZTR-OTHER—Maximum seizure tries field (other accesses).

OLCN—Overload class field (N = 0 to 15).

END—End indication field. Set to "1" to indicate the last word of the overhead message train; set to "0" if not last word.

RSVD—Reserved for future use, all bits must be set as indicated.

LOCAL CONTROL—May be set to any bit pattern.

P—Parity field.

☐ The registration ID message consists of one word. When sent, the message must be appended to a system parameter overhead message in addition to any global action messages.

T_1T_2 = 11	DCC	REGID	END	OHD = 000	P
2	2	20	1	3	12

The interpretation of the data fields is as follows:

T_1T_2—Type field. Set to "11" indicating overhead word.

DCC—Digital color code field.

OHD—Overhead message type field. Set to "000" indicating the registration ID message.

REGID—Registration ID field.

END—End indication field. Set to "1" to indicate last word of the overhead message train; set to "0" if not last word.

P—Parity field.

☐ The control-filler message consists of one word. It is sent whenever there is no other message to be sent on the forward control channel. It may be inserted between messages as well as between word blocks of a multiword message. The control-filler message is chosen so that when it is sent, the 11-bit word sync sequence (11100010010) will not appear in the message stream, independent of the busy-idle bit status.

The control-filler message is also used to specify a control mobile attenuation code (CMAC) for use by mobile stations accessing the system on the reverse control channel, and a wait-for-overhead-message bit (WFOM) indicating whether or not mobile stations must read an overhead message train before accessing the system.

T_1T_2 = 11	DCC	010111	CMAC	RSVD = 00	11	RSVD = 00	1	WFOM	1111	OHD = 001	P
2	2	6	3	2	2	2	1	1	4	3	12

The interpretation of the data fields is as follows:

T_1T_2—Type field. Set to "11" indicating overhead word.

DCC—Digital color code field.

CMAC—Control mobile attenuation field. Indicates the mobile station power level associated with the reverse control channel.

RSVD—Reserved for future use; all bits must be set as indicated.

WFOM—Wait-for-overhead-message field.

OHD—Overhead message type field. Set to "001" indicating the control-filler word.

P—Parity field.

Data Restrictions. The 11-bit word-sync sequence (11100010010) is shorter than the length of a word, and therefore can be embedded in a word. Normally, embedded word-sync will not cause a problem because the next word to be sent will not have the word-sync sequence embedded in it. There are, however, three cases in which the word-sync sequence may appear periodically in the FOCC stream. They are as follows:

the overhead message,
the control-filler message,
Mobile station control messages with pages to mobile stations with certain central office codes.

These three cases are handled by 1) restricting the overhead message transmission rate to about once per second, 2) designing the control-filler message to exclude the word-sync sequence, taking into account the various busy-idle bits, and 3) restricting the use of certain central office codes.

If the mobile station control message is examined with the MIN1 separated into NXX-X-XXX as described earlier (where NXX is the central office code, N represents a number from 2-9, and X represents a number from 0-9), Table 3-8 can be constructed to identify the central office codes which will cause the word-sync word to be sent. If a number of mobile stations are paged consecutively with the same central office code, mobile stations that are attempting to synchronize to the data stream may not be able to do

Table 3-11. Restricted Central Office Codes.

T_1T_2	DCC	Bit Pattern NXX	X	XXX	Central Office Code	Thousands Digit
01	11	000100(1)0000	...	...	175	0 to 9
01	11	000100(1)0001	...	...	176	0 to 9
01	11	000100(1)0010	...	...	177	0 to 9
01	11	000100(1)0011	...	...	178	0 to 9
01	11	000100(1)0100	...	...	179	0 to 9
01	11	000100(1)0101	...	...	170	0 to 9
01	11	000100(1)0110	...	...	181	0 to 9
01	11	000100(1)0111	...	...	182	0 to 9
0Z	11	100010(0)1000	...	...	663	0 to 9
0Z	11	100010(0)1001	...	...	664	0 to 9
0Z	11	100010(0)1010	...	...	665	0 to 9
0Z	11	100010(0)1011	...	...	666	0 to 9
0Z	Z1	110001(0)0100	...	...	899	0 to 9
0Z	Z1	110001(0)0101	...	...	800	0 to 9
0Z	ZZ	111000(1)0010	...	...	909	0 to 9
00	ZZ	011100(0)1001	0ZZZ	...	568	1 to 7
00	ZZ	111100(0)1001	0ZZZ	...	070	1 to 7
00	ZZ	001110(0)0100	10ZZ	...	339	8,9,0
00	ZZ	011110(0)0100	10ZZ	...	595	8,9,0
00	ZZ	101110(0)0100	10ZZ	...	851	8,9,0
00	ZZ	111110(0)0100	10ZZ	...	007	8,9,0
0Z	ZZ	000011(1)0100	0010	...	150	2
0Z	ZZ	000111(1)0001	0010	...	224	2
0Z	ZZ	001011(1)0001	0010	...	288	2
0Z	ZZ	001111(1)0001	0010	...	352	2
0Z	ZZ	010011(1)0001	0010	...	416	2
0Z	ZZ	010111(1)0001	0010	...	470	2
0Z	ZZ	011011(1)0001	0010	...	544	2
0Z	ZZ	011111(1)0001	0010	...	508	2
0Z	ZZ	100011(1)0001	0010	...	672	2
0Z	ZZ	100111(1)0001	0010	...	736	2
0Z	ZZ	101011(1)0001	0010	...	790	2
0Z	ZZ	101111(1)0001	0010	...	864	2
0Z	ZZ	110011(1)0001	0010	...	928	2
0Z	ZZ	110111(1)0001	0010	...	992	2
0Z	ZZ	111011(1)0001	0010	...	056	2
0Z	ZZ	111111(1)0001	0010	...	...	2

Note:
1) In each case, Z represents a bit that may be 1 or 0.
2) Some codes are not used as central office codes in the US at this time. They are included for completeness.
3) The bit in parentheses is the busy-idle bit.

so because of the presence of the false word sync sequence. Therefore, the combinations of central office codes and groups of line numbers appearing in Table 3-11 must not be used for mobile stations.

Forward Voice Channel

The forward voice channel (FVC) is a wideband data stream sent by the land station to the mobile station. This data stream must be generated at a 10 kilobit/second ±0.1 bit/second rate. Figure 3-4 depicts the format of the FVC data stream.

A 37-bit dotting sequence (1010...101) and an 11-bit word sync sequence (11100010010) are sent to permit mobile stations to achieve synchronization with the incoming data, except at the first repeat of the word, where the 101-bit dotting sequence is used. Each word contains 40 bits, including parity, and is repeated eleven times together with the 37-bit dotting and 11-bit word sync sequences; it is then referred to as a word block. A word block is formed by encoding the 28 content bits into a (40, 28) BCH code that has a distance of 5, (40, 28; 5). The left-most bit (i.e., earliest in time) shall be designated the most-significant bit. The 28 most-significant bits of the 40-bit field shall be the content bits. The generator polynominal is the same as that used for the forward control channel.

The mobile station control message is the only message transmitted over the forward voice channel. The mobile station control message consists of one word.

Mobile Station Control Message

2	2	2	9	5	3	5	12
$T_1 T_2$ = 10	SCC = 11	PSCC	RSVD = 000....0	LOCAL	ORDQ	ORDER	P
	SCC = 11		RSVD = 000...0	VMAC	CHAN		
2	2	2	8	3	11		12

The interpretation of the data fields is as follows:

$T_1 T_2$ —Type field. Set to "10."

SCC—SAT color code for new channel (see Table 3-7).

PSCC—Present SAT color code. Indicates the SAT color code associated with the present channel.

ORDER—Order field. Identifies the order type (see Table 3-8).

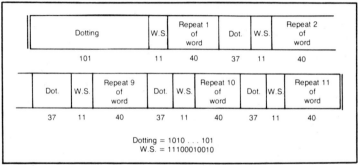

Fig. 3-4. Forward voice channel message stream.

ORDQ—Order qualifier field. Qualifies the order to a specific action (see Table 3-8).

LOCAL—Local Control field. This field is specific to each system. The ORDER field must be set to local control (see Table 3-8) for this field to be interpreted.

VMAC—Voice mobile attenuation code field. Indicates the mobile station power level associated with the designated voice channel.

CHAN—Channel number field. Indicates the designated voice channel.

RSVD—Reserved for future use; all bits must be set as indicated.

P—Parity field.

DEFINITIONS

Access Channel. A control channel used by a mobile station to access a system to obtain service.

Analog Color Code. An analog signal (see SAT) transmitted by a land station on a voice channel and used to detect capture of a mobile station by an interfering land station and/or the capture of a land station by an interfering mobile station.

BCH Code. Bose-Chaudhuri-Hocquenghem Code.

Busy-Idle Bits. The portion of the data stream transmitted by a land station on a forward control channel that is used to indicate the current busy-idle status of the corresponding reverse control channel.

Control Channel. A channel used for the transmission of digital control information from a land station to a mobile station or from a mobile station to a land station.

Digital Color Code (DCC). A digital signal transmitted by a land station on a forward control channel that is used to detect capture of a land station by an interfering mobile station.

Flash Request. A message sent on a voice channel from a mobile station to a land station indicating that a user desires to invoke special processing.

Forward Control Channel (FOCC). A control channel used from a land station to a mobile station.

Forward Voice Channel (FVC). A voice channel used from a land station to a mobile station.

Group Identification. A subset of the most-significant bits of the system identifications (SID) that is used to identify a group of cellular systems.

Handoff. The act of transferring a mobile station from one voice channel to another.

Home Mobile Station. A mobile station which operates in the cellular system from which service is subscribed.

Land Station. A station in the Domestic Public Cellular Radio Telecommunications Service, other than a mobile station, used for radio communications with mobile stations.

Mobile Identification Number (MIN). The 34-bit number which is a digital representation of the 10-digit directory telephone number assigned to a mobile station.

Mobile Station. A station in the Domestic Public Cellular Radio Telecommunications Service intended to be used while in motion or during halts at unspecified points. It is assumed that mobile stations include portable units (e.g., hand-held "personal" units) as well as units installed in vehicles.

Mobile Station Class. The following mobile station classes are defined for this section.

        Class I.   High power station.
        Class II. Mid-range power station.
        Class III. Low power station.

Numeric Information. Numeric information is used to describe the operation of the mobile station. The following subscripts are used to clarify the use of the numeric information:

    s to indicate a value stored in a mobile station's temporary memory.

    sv to indicate a stored value that varies as a mobile station processes various tasks.

    sl to indicate the stored limits on values that vary.

    r to indicate a value received by a mobile station over a forward control channel.

p to indicate a value set in a mobile station's permanent security and identification memory.

and

s-p to indicate a value stored in a mobile station's semi-permanent security and identification memory.

The numeric indicators are as follows:

$ACCOLC_p$. A four-bit number used to identify which overload class field controls access attempts.

BIS_s. Identifies whether a mobile station must check for an idle-to-busy transition on a reverse control channel when accessing a system.

$CCLIST_s$. The list of control channels to be scanned by a mobile station processing the Directed-Retry Task.

$CMAX_s$. The maximum number of channels to be scanned by a mobile station when accessing a system.

CPA_s. Identifies whether the access functions are combined with the paging functions on the same set of control channels.

DTX_s. Identifies whether the mobile station is permitted to use the discontinuous transmission mode on the voice channel.

E_s. The stored value of the E field sent on the forward control channel. E_s identifies whether a home mobile station must send only $MIN1_p$ or both $MIN1_p$ and $MIN2_p$ when accessing the system.

EX_p. Identifies whether home mobile stations must send $MIN1_p$ or both $MIN1_p$ and $MIN2_p$ when accessing the system. EX_p differs from E_s in that the information is stored in the mobile station's security and identification memory.

$FIRSTCHA_s$. The number of the first control channel used for accessing a system.

$FIRSTCHP_s$. The number of the first control channel used for paging mobile stations.

$LASTCHA_s$. The number of the last control channel used for accessing a system.

$LASTCHP_s$. The number of the last control channel used for paging mobile stations.

LT_s. Identifies whether the next access attempt is required to be the last try.

$MIN1_p$. The 24-bit number which corresponds to the 7-

digit directory telephone number assigned to a mobile station.

$MIN2_p$. The 10-bit number which corresponds to the 3-digit area code assigned to a mobile station.

$MAXBUSY_{sl}$. The maximum number of busy occurrences allowed on a reverse control channel.

$MAXSZTR_{sl}$. The maximum number of seizure attempts allowed on a reverse control channel.

N_s. The number of paging channels that a mobile station must scan.

$NBUSY_{sv}$. The number of times a mobile station attempts to seize a reverse control channel and finds the reverse control channel busy.

$NSZTR_{sv}$. The number of times a mobile station attempts to seize a reverse control channel and fails.

$NXTREG_{s-p}$. Identifies when a mobile station must make its next registration to a system.

PL_s. The mobile station rf power level.

R_s. Indicates whether registration is enabled or not.

RCF_s. Identifies whether the mobile station must read a control-filler message before accessing a system on a reverse control channel.

$REGID_s$. The stored value of the last registration number. ($REGID_r$) received on a forward control channel.

$REGINCR_s$. Identifies increments between registrations by a mobile station.

S_s. Identifies whether the mobile station must send its serial number when accessing a system.

SCC_s. A digital number which is stored and used to identify which SAT frequency a mobile station should be receiving.

SID_p. The home system identification stored in the mobile station's permanent security and identification memory.

SID_{s-p}. One of a number of system identifications stored in the mobile station's semipermanent security and identification memory.

SID_r. The system identification received on a forward control channel.

SID_s. The stored system identification.

$WFOM_s$. Identifies whether a mobile station must wait for an overhead message train before accessing a system on a reverse control channel.

Orders. The following orders can be sent to a mobile station from a land station:

>Alert. The alert order is used to inform the user that a call is being received.
>
>Audit. The audit order is used by a land station to determine whether the mobile station is active in the system.
>
>Change Power. The change power order is used by a land station to change the rf power level of a mobile station.
>
>Intercept. The intercept order is used to inform the user of a procedural error made in placing the call.
>
>Maintenance. The maintenance order is used by a land station to check the operation of a mobile station. All functions are similar to alert but the alerting device is not activated.
>
>Release. The release order is used to disconnect a call that is being established or is already established.
>
>Reorder. The reorder order is used to inform the user that all facilities are in use and the call should be placed again.
>
>Send Called-Address. The send called-address order is used to inform the mobile station that it must send a message to the land station with dialed-digit information.
>
>Stop Alert. The stop alert order is used to inform a mobile station that it must discontinue alerting the user.

Paging. The act of seeking a mobile station when an incoming call has been placed to it.

Paging Channel. A forward control channel which is used to page mobile stations and send orders.

Registration. The steps by which a mobile station identifies itself to a land station as being active in the system at the time the message is sent to the land station.

Release Request. A message sent from a mobile station to a land station indicating that the user desires to disconnect the call.

Reverse control channel (RECC). The control channel used from a mobile station to a land station.

Reverse Voice Channel (RVC). The voice channel used from a mobile station to a land station.

Roamer. A mobile station which operates in a cellular system other than the one from which service is subscribed.

Scan of Channels. The procedure by which the mobile station examines the signal strength of each forward control channel.

Seizure Precursor. The initial digital sequence transmitted by a mobile station to a land station on a reverse control channel.

Signaling Tone. A 10-kilohertz tone transmitted by a mobile station on a voice channel to: 1) confirm orders, 2) signal flash requests, and 3) signal release requests.

Status Information. The following status information is used in this section to describe mobile station operation:

> Serving-System Status. Indicates whether a mobile station is tuned to channels associated with System A or System B.
>
> First Registration ID Status. Indicates whether a mobile station has received a registration ID message since initialization.
>
> Local Control Status. Indicates whether a mobile station must respond to local control messages or not.
>
> Roam Status. Indicates whether a mobile station is in its home system or not.
>
> Termination Status. Indicates whether a mobile station must terminate the call when it is on a voice channel.

Supervisory Audio Tone (SAT). One of three tones in the 6-kilohertz region that are transmitted by a land station and transponded by a mobile station.

System Identification (SID). A digital identification associated with a cellular system; each system is assigned a unique number.

Voice Channel. A channel on which a voice conversation occurs and on which brief digital messages may be sent from a land station to a mobile station or from a mobile station to a land station.

Chapter 4

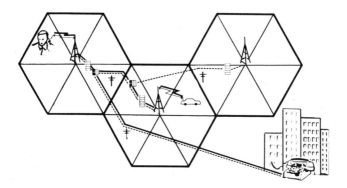

Available Systems and Equipment

Although this chapter won't cover every single piece cellular of gear that's ready for the marketplace, or even in the trial stage, what's contained will be pretty representative of those companies who can and will offer substantial contributions to this new avenue of communications. The companies mentioned in this chapter have given freely of their technical information and helpful applications. Their names will be around for a long time to come, and their contributions here merit close scrutiny when either buying or renting space and/or equipment. Their specifications are, indeed, excellent!

MOTOROLA'S DYNA T·A·C

Since Motorola was both willing and able to supply virtually full information on switch, mobile, and cell site systems, we are taking the liberty of giving Motorola precisely what it deserves - as much coverage as we can reasonably present to the reader for his edification and digestion. Therefore, this part of the book belongs to Motorola, along with our thanks for permission to use the material. DYNA T·A·C, of course, is Motorola's registered trade mark which it says represents in equipment and research $100 million and 1,000 man years of engineering to produce it. The system is already operating satisfactorily for American Radio Telephone Service in the Baltimore/Washington area and elsewhere.

Electronic Mobile Exchange

The Electronic Mobile Exchange (EMX) features digital switching along with stored program control available in central phone offices worldwide. Speech is digitized and all switching completed without blocking. Designed specifically for radiotelephone operations, EMX offers call waiting, no-answer transfer, call forwarding, and three-party conferencing, in addition to data on call recording transit information, hand-offs, subscriber options, and test call records. Call attempts, completions, Erlang traffic, and blocked calls are all recorded on 1,600 bpi 9-track tape with IBM standard labels. Figure 4-1 provides a look at the EMX. And with PCM (pulse code modulated) audio used within the EMX, reed relays, crossbars, or diode matrices are avoided as well as noise, distortion or crosstalk. Audio now appears a 4-wire with processing done via 24 channels in 1.544 Mbit channel banks. In addition, digitized audio from a special zone may be multiplexed to and from remote sites or directly from the telephone networks, if PCM trunks are available. A functional block diagram of the system is shown in Fig. 4-2.

Fig. 4-1. DYNA T·A·C system elements. (courtesy Motorola Communications)

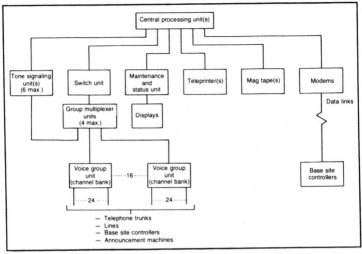

Fig. 4-2. EMX functional block diagram. (courtesy Motorola Communications)

Here you see the Central Processor (CPU), Switch Unit (SWU), Group Multiplexer Unit (GMU), Tone Signaling Unit (TSU), Maintenance and Status Unit (MSU) and, of course, inputs and outputs.

Central Processor (CPU). This CPU contains the switch controls, data acquisition, data base, cell coordination, fault management, and general call administration. Control and routing of calls are made within the CPU, while the switch unit manages timed audio cross-connections and dc supervisory signaling on the PCM buses. CPU nodes consist of a pair of microcomputers operating in *shared memory shadowing*, and each node communicates with other nodes through a parallel internodal link. This permits rapid and accurate processing of real-time events such as time changes, channel and line state changes. Per half node, the average number of instructions/second amounts to 400,000, with single bit parity checking on memory and peripherals. Analog phone signals received by EMX are converted to digital for distribution by various microprocessors (or microcomputers) distributed throughout the system. The CPU has a standby backup and either one or the other can continue full system operation if there are failures. Shared memory shadowing has both link control coupling between the twin microcomputers and a data coupling between nodes to maintain the necessary shared random access memory. Errors and faults have multidetection devices numbering six or more, including error counting and circuit removal from service.

Switch Unit (SWU). This switches PCM voice bytes, processes the PCM bit stream, and interfaces with CPU over data links. Operating independently of CPU except for updating, it controls digitized audio and tone signaling, and adds sync and signaling in the outgoing bit stream. Voice information proceeds to a time slot processor, and line signaling goes to the signal bit receiver which notifies the CPU's switch control unit of all changes in line signaling. A voice group contains 24 ports; (30 ports for international applications) each group multiplexer can interface up to 16 voice groups, or 384 voice facility ports; each EMX 500 can accommodate four group multiplexers, permitting 64 PCM ports. This amounts to 1,536 ports.

Sampling for the various switches occurs during 5-microsecond dwell times and audio becomes digitized and stored in PCM memory. Read ins and read outs then take place but readout is not sequential and the various bytes have been transferred to their respective lines of information, although bytes are written into odd PCM memory and read out of even PCM memory. So the connection of one telephone to another occurs as one legitimate number reaches control memory, is recognized and becomes linked with another as the various subgroups are accessed and results are written out of memories into output ports.

Group Multiplexer Unit (GMU). Group Multiplexers convert serial bipolar bit streams into 8-bit parallel format and receive time multiplexed 8-bit parallel words from the switch unit, converting this to serial readout, and transmit resulting PCM bit streams to the voice group units (VGUs), along with parallel interfaces to the TSU Tone Signaling Unit. The GMU also time multiplexes the 8-bit parallel format in the group interfaces to the SWU. With differential transmission paths between the GMU and the SWU removing most or all logic ground noise, trunk, timing, and signaling/supervisory tones all travel through SWU ports. Telephone trunks and radio channels may be multiplexed, thereby interleaving these across several VGUs.

Should one VGU fail, complete switching may continue but with reduced system response. The EMX is not equipped with the channel banks that perform A/D and D/A conversion and time division multiplexing of 24 analog ports for bit stream interchange with the EMX. They may be either customer or Motorola supplied as optional equipment with independent power and alarms and are to be located within 250 cable meters of the EMX unless there are repeaters.

Tone Signaling Unit (TSU). Voice frequency signaling tones and all supervisory tones are generated or decoded by the TSU via PCM digitized format. Digital lookup tables or Read-Only-Memory (ROM) contains stored data for all signaling and supervisory tones, thereby eliminating either frequency or level adjustments from all points other than the external interface card. Receive and transmit PCM, in addition to tone generator sync, are all illustrated in Fig. 4-3. It shows the tone bus for the various receivers and their tone data inputs to the interface above.

Maintenance and Status Unit (MSU). The entire EMX and the MSU, itself, are constantly checked by this unit for faults to prevent system shutdown. And when a failure appears, the Central Processor (CPU) finds the problem circuit or module and isolates it from the rest. When the MSU notes a failure, it strobes both the Alarm and Status Panel (ASP), while base site controllers' status is reported also. Figure 4-4 illustrates both SSR and MPROC processor and memories and the various buses leading to them. Each ASP contains up to eight alarm and status modules and is controlled by the active MSU processor and then interfaces with the CPU.

I/O Hardware. Teleprinter equipment has to conform to Bell

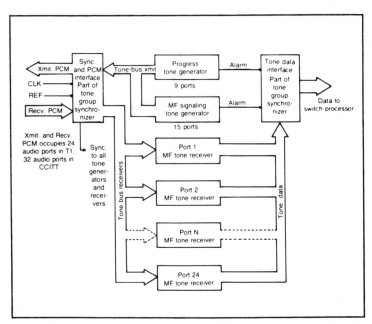

Fig. 4-3. Typical tone signaling unit (TSU) block diagram. (courtesy Motorola Communications)

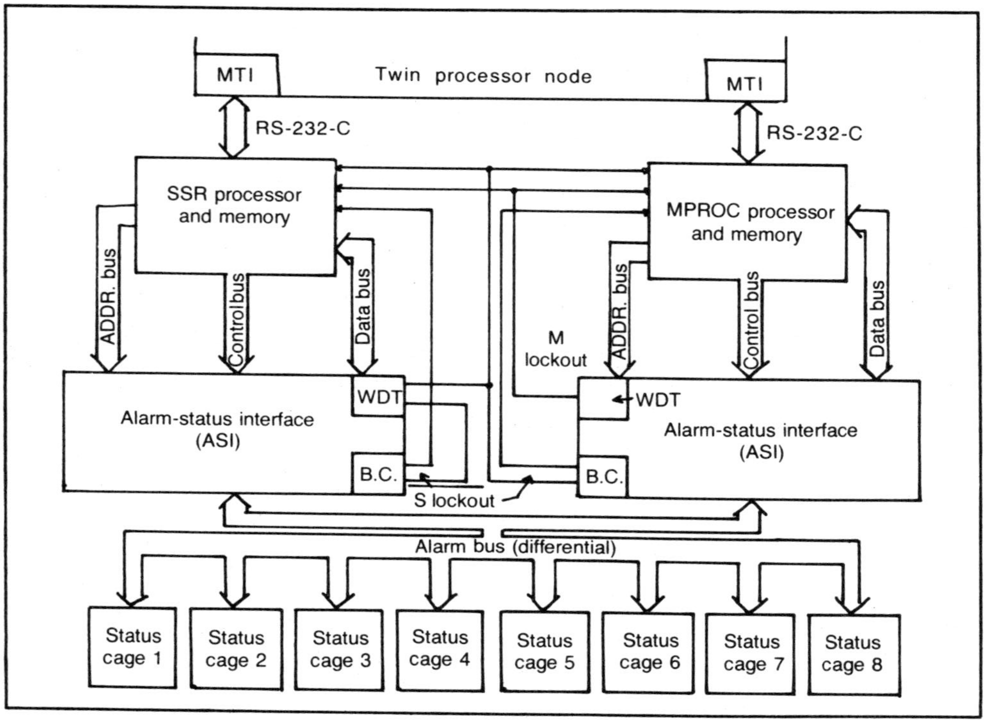

Fig. 4-4. Maintenance and status unit (MSU) simplified functional diagram. (courtesy Motorola Communications)

System Technical Reference PUB 41715, *Data Speed 40 Stations for Dataphone Service*, dated December 1973. Any EMX must have one teleprinter, and the minimum number of teleprinter ports equipped for a redundant system is six. The maximum number of teleprinters possible is 14. Magnetic tape has to meet ANSI standard X3.39 (1973) for ID bursts, interblock gaps, file marks, etc., conforming to *Recorded Magnetic Tape for Information Interchange (1600 CPI Phase Encoded)*. Alarm and status display panel offers visual indicators of alarm and/or status of the EMX control system and radio channels. Each ASP has an alarm and status electronic board and is managed by the MSU processor.

Software. Motorola calls EMX software "a collection of processes which are activated in response to messages." And a process is said to be a high level function, including one or more programs in node memory that will make them execute.

There are single task processes, clocked task processes, and sequential processes, and they must be completely executed before others may be begun. They can be hardware interrupts, line changes, and node commands; and tasks may communicate with one another directly, but with other processes only through messages. All tasks have specifically labeled entry and exit points. Such processes and tasks may be executed concurrently in the EMX and are controlled by an Executive much like conventional computer programs. It recognizes four types of tasks:

☐ Interrupt response: to run when there are hardware errors or triggered by a timing device.

☐ Clocked task: operate between 10 msec and 1.27 seconds in 10 msec increments.

☐ Scheduled sequential processes: in response to message inputs from other jobs, state changes, and timeouts.

☐ Scheduled single tasks: run when responding to other job messages.

Motorola calls an individual process that is in progress a job. And some node has been assigned this job number by the Executive. The job also has assigned a memory block for data in sequential operations. With job memory blocks, the same process may be commenced and executed at several locations within the EMX as needed.

An EMX database consists of files containing a number of varieties of tables. There are state, time, and task tables, as well as dynamic information such as time and line states or constant values "such as class codes and process identification numbers." Stored are subscriber and equipment data, numbering plans, add, change, or delete orders, or access commands. As you can understand, then, every node has its own program and individual memory, although there may be memory transfer between two nodes. An Executive is loaded into each node during initial programming; as messages come in, the Executive transfers control to the indicated programs which perform data handling, message generation, or communications routines.

Firmware. Firmware is another operational level in EMX that has both software and hardware appearances. It consists of stored programs in the ROM or PROM and usually does signal translation, small computations, and message generation. Ordinarily, such programs are written in the Motorola EMX development center and loaded into microprocessor chips for eventual execution. They are identified as follows:

☐ MPROC, SSR: alarm and status display and Watchdog Timer refresh requirement messages.

☐ TCB: tape drive messages in real time.

☐ SCI: signal bit changes and switch alarm reports.

☐ SCIP: serial interface peripheral.

There is a great deal more to this software/firmware procedure, but it's all laid out in Motorola's Cellular Radio Telephone Systems manual for interested people to read. As is, you have enough to work on here, not counting the system diagnostics and maintenance information contained in the last chapter. So let's get along to the rest of the system before you become completely inundated with computerized detail.

Base Station Equipment

Motorola's DYNA T·A·C, base stations shown in Fig. 4-5, are available in two versions for 800 MHz cellular radios. The 8-channel unit has the following:

☐ A receive bay with 8 voice channels, signaling and scanning receiver.

☐ A power supply bay furnishing one charger for 8 voice channels.

☐ A transmitter bay with 8 voice channels, a stable frequency source, and a signaling transmitter.

The 16-channel unit has the following:

☐ A receive bay with 16 voice channels, signaling and scanning receiver.

☐ A power supply bay furnishing two chargers for 16 voice

Fig. 4-5. DYNA T·A·C base station. (courtesy Motorola Communications)

channels, Two transmitter bays with 16 voice channels, a stable frequency source, and a signaling transmitter.

A functional block diagram of both the receiver and the transmitter is shown in Fig. 4-6 and Fig. 4-7. Note especially the UHF splitters and matrices in the receive bay, and the modulation arrangements following the distribution amplifier in the transmitter. Harmonic filters to reduce spurs and secondary emissions are always added to transmit outputs.

Broadband preamplifier/mixers convert antenna inputs to i-f frequencies in sector receive as a UHF splitter separates voice and signaling information for the first mixer and scanning receiver. A 15

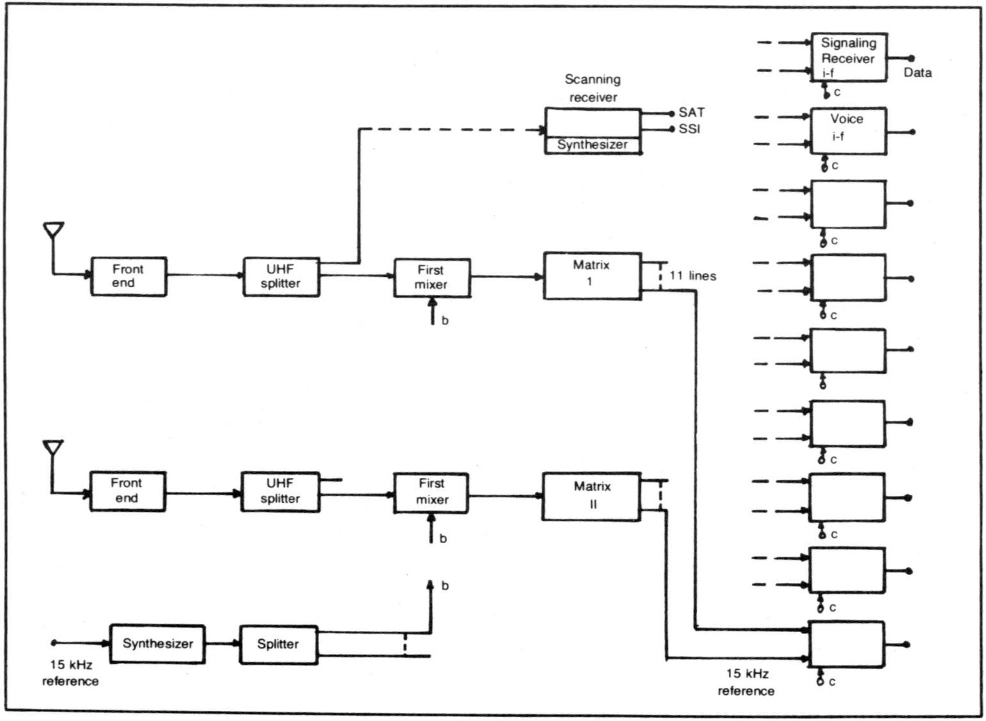

Fig. 4-6. Receive bay functional block diagram. (courtesy Motorola Communications)

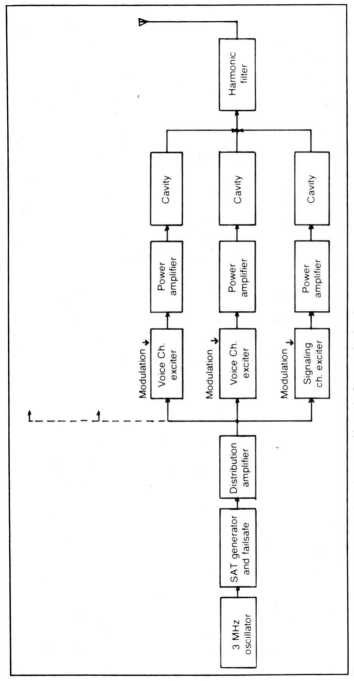

Fig. 4-7. DYNA T·A·C base transmitter. (courtesy Motorola Communications)

kHz reference generated in the synthesizer also reaches the first mixer as well as signaling receiver i-fs which output data flow. A commutating switch steps the scanning receiver through its six sector antenna inputs. Each front end serves 16 voice channels and identical i-f modules process each voice channel.

Featured in the system is Motorola's maximal ratio 2-branch diversity combiner, which is so designed that all sensitivity and selectivity characteristics are known before they arrive at the combiner. The 2-branch diversity combiner is a true maximal ratio device rather than one of equal gain, by reason of its wide dynamic square law mixing technique rather than a traditional fed-forward, variable gain amplifier for each branch. Here, signals are summed according to their individual signal strengths, rather than according to some arbitrarily set gain. This approach contributes to improved signal-to-noise ratios. Regardless, all summed branch signals are made available to the receiver's output. Diversity techniques not only offer additional adjacent channel rejection, but they also improve cochannel interference as well, since average signals are maintained at higher levels, thereby boosting the carrier-to-interference (C/I) ratios. Redundant signaling channel receivers and scanning receivers are also available as options. Rf sensitivity is specified at −116 dBm and spurs and image rejection at 80 dB.

Transmit bays have a 3 MHz oscillator with frequency stability of 1 ppm/year, and output power of 20 watts/channel. In each bay there are 8 voice and 1 signaling channels, with minimum channel spacing of 630 kHz. SAT tones and failsafe alarm frequencies proceed through a distribution amplifier to the various voice exciters where modulation is introduced. The combined signals are then power amplified and put into a cavity combiner with verifying channel feedback. Whenever some signaling channel, for any reason, cannot be combined with voice channels, its output is connected to a separate antenna. The second transmit bay in any basic group is identical to the first, but does not have the reference oscillator or a signaling channel unless optionally equipped. Spurs and harmonics are more than 70 dB down from nominal carrier frequencies, and sideband noise is more than 80 dB down.

Mobiles and Portables

Two types of radios have been designed especially for cellular operations: handhelds (portables) and mobiles. Both are transceivers, but one operates on very small power while the other has an output approaching that of AM CB. They cover the full 666 rf

Fig. 4-8. DYNA T·A·C portable radiotelephone. (courtesy Motorola Communications)

channels in full duplex (825.030 to 844.980 MHz in transmit, and 870.030 to 889.980 MHz in receive) at maximum power outputs of 0.6 watt and 3 watts, respectively. Channel spacing for each is 30 kHz, with a frequency stability of ±2.5 ppm (parts/million), or 0.00025 percent.

Mobiles. The mobile is a full duplex, state-of-the-art modular radiotelephone based on the Pulsar mobile phone family and other Motorola 2-way radio experience which includes LSI, crystal filters, and frequency synthesizers. Low loss duplexers, a microstrip power amplifier section with power output leveling, along with modular construction for easy circuit board exchange, and a cast chassis with thick top and bottom covers should combine to make this a very serviceable and rugged unit.

The mobile control assembly, fabricated of high impact plastic, offers contemporary styling, clearly visible graphics, including numbered pushbuttons, all of which are illuminated at night. Motorola lists its features as speaker phone, recall of last number dialed, function status display, pushbutton dialing, electronic

Fig. 4-9. DYNA T·A·C mobile radiotelephone. (courtesy Motorola Communications)

scratchpad memory, DTMF end/end signaling, dialed number display, electronic lock, service mode provisions, abbreviated dialing, volume controls, auxiliary call alert, and on/off-hook call placement.

In VSP operation, the control unit uses a separate loudspeaker and microphone as long as the handset is at rest. Picking up the handset will disable the speaker phone and resume usual handset operation. On the back of the handset is a pushbutton keypad, positioned so the user can hold on and select any phone number he desires with one hand. As numbers are pushed, audible tones feed back, indicating contact. Also, all numbers are displayed to be checked for selection error. Ten most frequently called numbers may also be stored in memory for quick recall, but may be changed or deleted at any time by the subscriber. Their retention is protected by a battery.

An electronic scratchpad and last number recall are also offered. Phone numbers noted during any conversation may be entered into the mobile's temporary memory from which it may be dialed by single-button impress. Such numbers may also be loaded into other more permanent dialing memories if desired. Then, if some number you wish to reach is busy, the mobile's control unit supplies last number recall, by single button transmit from the temporary storage register. This number remains until replaced by a second keypad entry. To unlock the mobile, a 3-digit security code must be entered after initial power up.

Motorola tells me that all dialing operations are completed before the call, and numbers are not outpulsed until the appropriate time.

When a number is dialed and the transceiver is in contact, the control unit displays such functions as in use, no service, roam, lock and auxiliary alert. This alert operation may also trigger some external alarm such as truck/auto horn or headlights when there are incoming calls. Under normal circumstances. an electronic ringer notifies the subscriber of any receiver call.

Maintenance for mobiles includes remote access to key test points and transceiver exercise routines to establish fault locations. A single external connector for such test points is provided. The radio unit measures 2.5 × 10 × 11.75 inches, and the control unit 2.8 × 4.1 × 8.9 inches, with the former weighing 13.7 lbs. and the latter 1.8 lbs. Supply voltage is specified at 13.7 V, negative ground, with current drain of 3 amps in transmit and 0.8 amps in standby. The FCC type acceptance number is ABZ89FT5612.

Portables. Undoubtedly, the portable (hand-held) unit has the same servicing provisions, plus a 12-button telephone format keypad, the usual phone number display, recall of the last number dialed, and the 10 most frequently called phone numbers may be stored in memory for quick recall. Earpiece volume and received call rings are independently controlled by a pair of keypad buttons, with last level setting retained until levels are re-adjusted. Full duplex operation continues, and there is also a Voice Operated Transmission mode (VOX) included to conserve battery drain and run the transmitter only during actual conversation. Motorola lists portable features as full duplex operation, VOX, pushbutton dialing, dialed number display, abbreviated dialing, recall of last number dialed, electronic scratchpad memory, electronic lock, electronic volume controls, functional status display, DTMF end-to-end signaling, and on-hook dialing.

Thick and thin film hybrid circuit modules are used throughout as well as custom linear and CMOS integrated circuits. Transceiver call processing, signaling and logic operations are controlled by a CMOS microprocessor.

The portables operate on supply voltages of 7.5 V between −30°C and +60°C, and are specified to supply 12 three-minute conversations over eight hours of operation on an overnight charge. They are 8.1 inches long, 1.8 inches wide, and 2.8 inches deep and weigh 1.8 lbs.

Spurs are −60 dB below carrier. FM hum and noise measure −32 dB below a 1,000 Hz tone for ±8 kHz deviation. In receive, sensitivity is reported at less than 0.35 μV (12 dB SINAD; C message) and selectivity of 65 dB audio distortion at less than 5 percent. Spurious radiation should be held to −89 dBm in the receive band.

Portables have both adapter accessories for standard usage and cigarette lighter/ac charging.

OKI'S CONTROL AND TRANSCEIVER UNITS

One of only several who showed willingness to cooperate because of unusually intense competition, OKI Advanced Communications, Hackensack, New Jersey, courteously supplied preliminary information on both its UM 1043 mobile transceiver and RZ 4205 control unit. The ensuing discussion, therefore, is based on these specifications, with all diagrams and photographs supplied directly by OKI. Because of on-going technical development such specifications may change with or without notice, according to that manufacturer's objectives and interests. I report, consequently, on such details as are immediately available.

Most descriptions begin with the control unit and work through the transceiver. Much in the way of controls, however, have already been described in other parts of the book, so I'll arbitrarily begin with the transceiver, work through controls, and then join the two in a diagram that should put the entire mobile together in a relatively easily understood package.

The Transceiver Transmitter

Designed for normal 825-845 MHz mobile transmit and 870-890 mobile receive bands, this transceiver, by the addition of another duplexer module, can also accommodate 998 channels rather than the 666 channel pairs presently specified by the Federal Communications Commission. These reserve frequencies lie between 821-825 MHz, 845-851 MHz, 866-870 MHz and 890-896 MHz, respectively, for transmit and receive, and could be added in time due to the huge cellular market anticipated as equipment and service prices decrease.

An appropriate diagram illustrating the two transceivers is shown in Fig. 4-10. As indicated, these are modular transceivers, with each module separate from the others, requiring no solder connections when replacing. As you can see, the difference between the 666-channel and 998-channel systems is the addition of an amplifier and dielectric receive bandpass filter between the receive dielectric duplexer and the radio module. Otherwise, the two systems appear similar in block diagram form. Only replacement of the duplexer module is affected. The UM 1043 has been designed to accommodate either system.

In the transceiver are six major modules which are now identified along with their functions:

☐ The *Radio/Audio Module* generates an 800 MHz +20 dBm modulated carrier; demodulates rf signal receive carrier; processes transmit and receive audio; operates receive local oscillator and transmit carrier signals; and includes audio filtering, companding, mute switches and deviation limiter.

☐ The *Power Amplifier Module* amplifies + 20 dBm carrier to 3 watts and exercises power control and isolation.

☐ The *Duplexer Module* has transmit bandpass filter, harmonic filter, and duplexer and operates rf preselector.

☐ The *Logic Module* executes data transmission and reception, SAT detection/transponding, measures received rf signals,

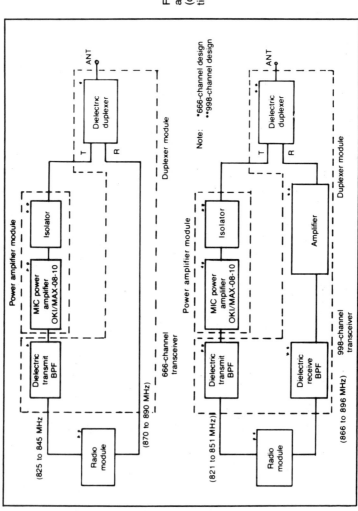

Fig. 4-10. Block diagram of the 666- and 998-channel transceiver units. (courtesy Advanced Communications)

call processing logic operations, and includes DTMF and audio tone generator.

☐ The *Power Supply/Interconnect Module* supplies regulated voltages, offers supply protection, furnishes interconnects between modules, and a receptable for NAM socket.

☐ The *Main Frame Module* houses other modules, furnishes control, antenna and battery interfaces, and contains serial number unit. The transceiver also has an 8-line bidirectional data bus, permitting direct hookup of test equipment for checks and troubleshooting. Five of these lines interface with the control unit. The transceiver also has a 36-point connector to identify and isolate any and all faulty modules located internally.

Transceiver features include auto levelling of transmitted power output by cell site in eight 4 dB power step attenuations; a temperature-compensated crystal oscillator offers stability of ±2.5 ppm per year within ambient temperatures from −20°C to +50°C; a dual conversion receiver with 45/10.7 MHz i-fs provides at least 12 dB SINAD sensitivity at −116 dBm; transmitter keying protected when hardware or software errors appear; final power amplifier fully protected from improper antenna termination; high and low voltage shutdown protection circuits; transceiver disabled with thermal shutdown when temperatures rise above normal; signal-to-noise ratios are aided by compandor circuits; dual-tone, multifrequency signals may be generated, offering end-to-end signaling; backup standby battery in transceiver to maintain memory banks during receiver off times.

The transmitter provides eight output power levels, with a nominal unattended power level output of 3 watts into a 50 Ω load. As controlled by binary signals from the logic circuits, rf power levels are as follows in dB: 0, −4, −8, −12, −16, −20, −24, and −28, and remain within ±2 dB of any of these levels. At 3 W, the nominal power amounts to 34.8 dBm. Time required to change rf output from one value to another is within 2 milliseconds.

When the carrier-on state is inactive, transmitted rf power output does not exceed −60 dBm. An rf power detector continuously monitors the transmitter and generates an rf power-on signal when the power output level meets or exceeds the threshold between −60 and +4.8 dBm, with response time not in excess of 100 msec.

Modulation signals may be produced with a peak deviation of ±12 kHz, when modulated by a 1 kHz tone. And when the 1 kHz tone generates peak deviation to a maximum of ±8 kHz, deviation remains within ±10 percent over the initial value over the specified temperature range. Maximum modulation sensitivity change following channel switching does not exceed 10 percent under all ambient (environmental) conditions. Voice, wideband data, SAT and signaling tones are included in transmitter inputs.

Voice circuits (Fig. 4-11) consist of a line receiver, bandpass filter, compressor, preemphasis circuit, deviation limiter, low pass filter, audio muting, and summing network. Here signaling and tone signals, in addition to DTMF and wideband data, may be added before all information reaches the rf modulator for antenna transmit.

The compressor has a 2:1 compandor; so for every 2 dB input change, the output change level is 1 dB and modulation plus predistortion for the complementary expander found in the cell site receiver. The compressor features an attack time of 3 msec and a recovery period of 13.5 msec, each ±20 percent, as required.

Before the compressor is an audio bandpass filter having attenuation that increases at least 24 dB/octave under and above 0.3 kHz and 3 kHz, respectively. Preemphasis rates 6 dB/octave between the same limits, and the following low-pass filter attenuates at no less than 36 dB/octave above 3 kHz and at least 35 dB/octave above 5.9 kHz. Audio muting will insert at least a 40 dB loss in the voice path, with mute transition times less than 10 microseconds.

Wideband data input relies on Manchester encoding, transforming each non-return-to-zero (NRZ) binary one to a 0-1 transition and each NRZ binary zero to a 1-0 transition. This permits 10 kilobit BCH encoded data to be transformed to a 20 kilobaud data rate. Binary frequency shift keying with peak deviation of ±8 kHz is used. Nominal passband at 3 dB down for wideband data input to the

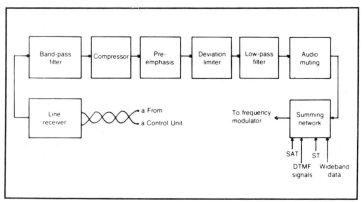

Fig. 4-11. Transmit audio processing. (courtesy Advanced Communications)

frequency modulator is between 200 Hz and 20 kHz, with rolloff below 200 Hz of 6 dB/octave and above with at least 38 dB of attenuation past 60 kHz. FM sensitivity remains constant within ±10 percent over the given passband.

Dual tone multifrequency signaling (DTMF) is derived from a pair of sinusoidal high and low frequencies consisting of 1.209, 1.336, and 1.477 kHz in one group, and 609, 770, 852, and 941 Hz in the other. Each key between the wire symbol (#) and digits 0 through 9 makes use of two of these assigned frequencies when pressed on the keyboard. All 7-tone pairs are stabilized within ±1.5 percent of given values. DTMF sidetones returning to the receive-audio lines at the control/transceiver unit have a level of −20 dBm ±3 dB with a 600 Ω termination. Voice/noise on the receive-audio line are simultaneously attenuated by at least 45 dB. End-to-end signaling may either be transmitted manually or stored and automatically transmitted when required. This occurs when user dialing occurs at a rapid rate so that digital information becomes stored until there are sufficient time intervals to allow accurate pulse transmissions.

SAT (Supervisory Audio) Tones at 5.97, 6, and 6.03 kHz are included within phase specifications of 0 ± 20 degrees, and phase step responses settle within 10 percent of final steady state by or before 250 msec.

ST (Signaling) Tone to the cell site has a 10 kHz ± 1 Hz tolerance as specified, with total harmonic distortion of 20 dB or more below the fundamental. Peak deviation under modulation is ± 8 kHz, ± 10 percent. With carrier on and transmit-audio muted, FM hum and noise are at least 32 dB below a 1 kHz signaling tone. Residual AM is 5 percent less than the transmitter carrier voltage, with any rms frequency deviation due to audio distortion at least 26 dB below this same tone level.

Emissions between 870 and 890 MHz on 30 kHz center (mean power with modulated carrier) do not exceed −120 dBm, and harmonics and spurs are at least 60 dB below the transmitted carrier.

Transceiver Receiver

Receivers operate within the 870 to 890 MHz band and will easily receive cell site transmissions with frequency tolerances of + 1.5 ppm and deviations up to + 12 kHz. Baseband outputs are available for voice, wideband data, and Supervisory Audio Tones (SAT). Voice receive circuits following the FM demodulator (Fig. 4-12) are made up of a mute switch, de-emphasis and bandpass

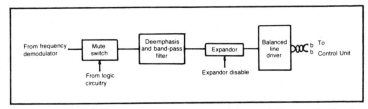

Fig. 4-12. Receive audio processing. (courtesy Advanced Communications)

filter, the expandor and balanced line driver that services the control unit.

The Mute Switch can insert no less than 40 dB in the audio with transition times of less than 10 microseconds, and switch voltage transients do not exceed 5 percent peak value of the 1 kHz tone at 8 kHz deviation.

The Expandor/Compandor reverses the transmitter compression ratio, and for every 1 dB input level change, produces a 2 dB output level expansion from either a companion mobile or the cell site transmitter. Expandor and compressor attack and recovery times are, of course, identical. Normal expandor input voltage is equivalent to a 1 kHz tone from a carrier with ± 2.9 kHz peak frequency deviation, having an output tolerance of ± 1 dB over inputs of +15 dB to −21 dB at 200 Hz to 3,500 Hz.

Receiver distortion with a 1 kHz tone at −50 dBm carrier is at least 26 dB below the 1 kHz demodulated tone, and hum and noise are 32 dB or more down from the modulated carrier at the same sensitivity. The frequency discriminator will handle either wideband FSK or SAT to the logic section, and at room temperature that's a flat ± 2 dB from 0.2 kHz to 20 kHz, with rolloff above and below these frequencies that is not less than 6 dB/octave. Output from the discriminator is normally within ±10 percent of modulation in the band. Incoming Manchester-encoded wideband data decodes into NRZ format, and bit-word sync is recovered, and so is SAT. Receiver emissions within the receive band do not exceed −80 dBm.

Receiver sensitivity begins at −116 dBm and will settle within 10 percent of its final value within 500 microseconds, ±20 percent for step changes of ± 20 dB within the specified range. Predetection bandwidth measures no greater than 32 kHz at the 6 dB down points, and intermodulation distortion resulting from unmodulated rf signals of 60 and 120 kHz at −35 dBm does not exceed −100 dBm. And SINAD with expandor disabled, does not decrease more than 6 dB with interfering signals at 1 kHz. For interfering signals below

854 MHz or above 908 MHz, performance "is met with the interfering signal level 100 dB greater than the on-channel signal."

Audio tones consists of two sine waves of 770 Hz ± 40 Hz and 1,150 Hz ± 55 Hz that are used to indicate a number of equipment conditions to the operator and may be injected into the receive-audio line of the Control/Transceiver unit as controlled binary signals. Signal levels are −22 dBV across a 600 Ω termination. Voice or noise energy appearing on the receive-audio line is simultaneously muted by at least 45 dB.

The Control Unit

This unit is the RZ 4205 control for UM 1043 transceiver, usually mounted in the mobile within reach of its operator. It includes a handset, keypad, loudspeaker, all mobile controls, visual indicators, and a dialing number display visible to the operator when receiving or initiating calls. A loudspeaker in the control unit alerts the operator for incoming calls and provides audible confirmation of call functions when placing calls with the handset cradled. Controls, indicators, and keyboard functions are illustrated in Fig. 4-13. A short explanation of these features follows:

Pushbutton dialing permits phone numbers to be entered just as in an ordinary home set. Numbers are dialed either on- or off-hook, according to user preference. When the Send key is pressed, dialed numbers stored in memory are transmitted, with the last number retained in memory and automatically redialed by again engaging the Send key if it is busy or there is no answer. Call progress may be monitored without unhooking the phone until the called party answers. Ten phone numbers of up to 16 digits each may be stored in memory, which is user programmable via the Store key. Any stored number may be recalled by pressing the Recall key followed by the address. A clear 7-digit LED display shows the number being dialed and the most recently entered seven digit number appears in the display. A Nite key dims the display when used at night, while all keys and labels are back lighted for poorly illuminated conditions. An electronic lock, activated by the Lock key, prevents unauthorized use of the equipment without knowing the combination of three or four digits entered by an operator at the time of installation. A positive latch secures the handset to prevent it falling free with vehicle motion.

Other features include an audio control that will insert 30 dB of attenuation, a handset volume control that delivers 20 dB of attenuation, and an alert volume control which offers 0, 10, and 29 dB

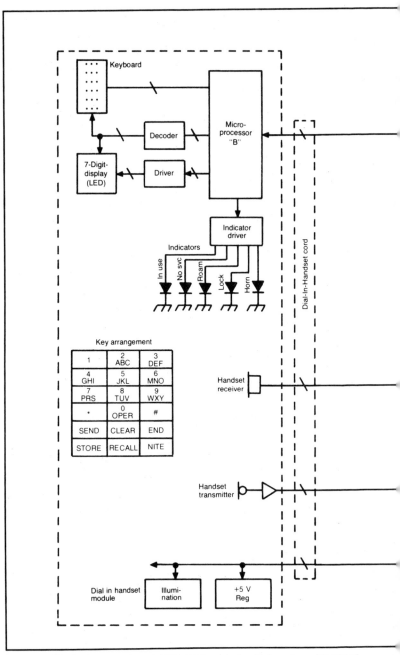

Fig. 4-14. Block diagram of model RZ 4205 control unit. (courtesy Advanced Communications)

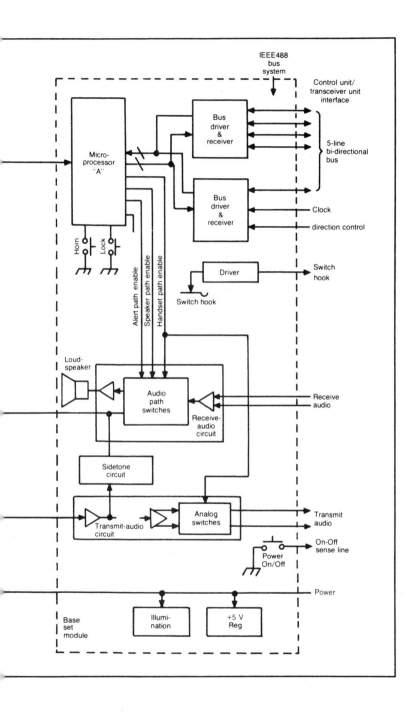

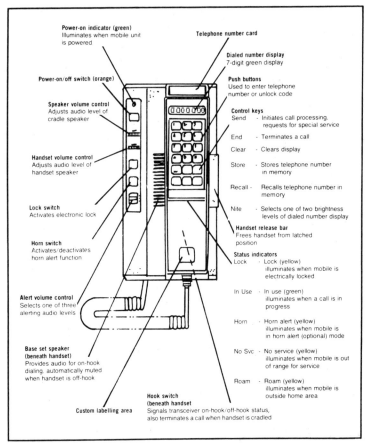

Fig. 4-13. Controls and functions—Model RZ 4205 Control Unit. (courtesy Advanced Communications)

attenuation. Mobile equipment status is indicated by In Use, No Service, Roam, Lock, and Horn. The horn alert will turn on a vehicle's horn or headlights if there is an incoming cellular call when unattended. The block diagram of the RZ 4205 control unit in Fig. 4-14 shows microprocessors, indicators, audio paths, sidetones and several signal, voltage, and control paths. Approximate weight on the control unit is 1.8 lbs. To protect against possible ear drum damage a limiter in the handset shuts down audio after 120 dB_{spl}. SPL amounts to 20 log P/Po, where P is the rms sound pressure in Pascals and Po = 2×10^{-5} Pascal. And an input level of 97 dB_{spl} results in an output level of -20 ± 3 dBV at 1 kHz in handset excitations. For receive, a 1 kHz -20 dBV input produces an output of 94 ± 3 dB_{spl} with volume set for maximum. Sidetone signal response for a 1 kHz of 97 dB_{spl} results in a 85 ± 3.5 dB_{spl} at the artificial ear. As in receive, the same input generates an output of 81 ± 6 dB_{spl} with the speaker or alert volume control set for an attenuation of 10 dB.

The Interface

As promised at the beginning of the chapter, Fig. 4-15 shows the interface between control unit and transceiver in a broadly illustrative diagram. Battery inputs, vehicle ground and control contacts are included, and descriptive blocks in the control unit should aid in understanding the operation of this portion also.

I hope this discussion, although containing some rather dry statistics, has added another notch in accumulating knowledge of cellular operations. At this stage, information either hasn't yet been fully developed or is closely guarded before final release.

ANACONDA-ERICSSON SYSTEM

Jointly owned by Atlantic Richfield and LM Ericsson Telephone Co. of Stockholm, Sweden under the name Anaconda-Ericsson, cellular products from this enterprise will surely come to market, backed by over 100 years of telephone experience from the parent Swedish company.

Offered as the CMS 8800 system, it includes a digital AXE 10 switch and F-900 radio base units. According to Ericsson, any mobile will work with the system "provided it fulfills FCC specifications for cellular mobile radio."

Based on the Nordic Mobile Telephone System (NMT) in Denmark, Finland, Norway, and Sweden and activated in 1981, this automatic service is designed to completely replace various manual systems with a 450 MHz service band for all four countries. Tested since 1977, it is similar in many respects to that authorized by our own FCC but slightly more sophisticated due to an all-digital switch developed expressly for cellular radio rather than a standard telephone modification. Base and mobiles in NMT have transmitter outputs of 50 and 15 watts, respectively. The new 8800, however, is wholly compatible with U.S. types including the frequency range between 820 MHz and 900 MHz. Ericsson says that more than 200 of its digital switches are operational in over 23 countries.

The CMS 8800 System

The CMS 8800 evolves about a network of radio base stations,

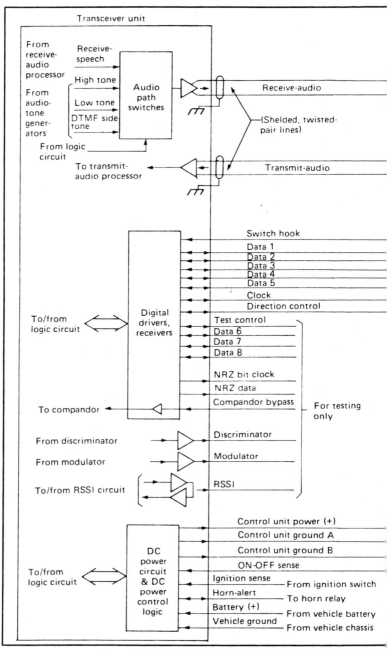

Fig. 4-15. Control unit/transceiver unit interface. (courtesy Advanced Communications)

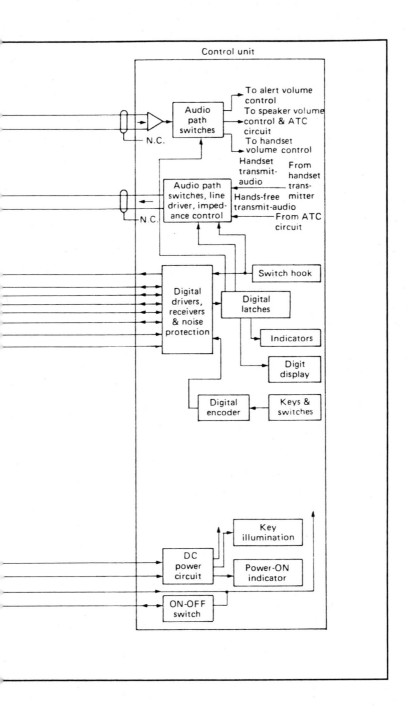

connected to a fixed telephone network by way of telephone exchange provided for mobile telephone (MTX). Connections to the phone network may be at the toll-tandem-transit level or it may be used as PABX to some local exchange. Base stations connected to the MTX constitute a service area, with selective calling to mobiles transmitted in parallel throughout the area. There may also be traffic areas offered within a single service area and directly addressed by the base stations involved.

When receiving a mobile call from another distant subscriber, the MTX checks for a valid number, the subscriber category and whether this individual is within the immediate service or traffic areas. Should these conditions be met, the MTX transmits a selective signal on all calling channels in the area. An idle mobile will immediately pick up this call and "home in" on the particular frequency. An automatic acknowledgment from the mobile follows, locating its position and thereby that of the nearest base station. The MTX seizes an available channel on the particular base station, the mobile is informed, and normal handshaking takes place to secure a synchronized connection.

When originating a call, the desired number is dialed, a calling channel seized, followed by handshaking, and eventual verification and transmission into the telephone network via MTX.

Calls in progress are supervised by a SAT tone signal about 6 kHz which is constantly evaluated by the base station. If the signal-to-noise ratio becomes weak, the MTX—which has been monitoring adjacent receivers—determines if another base station might be profitably used instead. Should this be the case, MTX switches the mobile to the new channel and conversation or data flow continues. When all channels are occupied, no action occurs, but this is normally only a problem during highest peak usage.

Roaming

Roaming, a difficult problem not yet fully characterized in the U.S., has one home base telephone exchange (MTX). When a mobile is operating in a distant service area recognition is automatic after it contacts the new MTX and is registered accordingly as being in a foreign area. Calls from home base are then rerouted to the new telephone exchange. When the mobile returns to its home area, the MTX calls the foreign MTX and the mobile's signs are erased from the distant area and once again stored for local action.

At least that's the way it's temporarily set up so that calling may be automatically processed from valid subscribers, and completed either through a mobile operator, or using end-to-end access. The CMS 8800 system may also accommodate other CMS 8800 systems in automatic roaming, and roaming via other foreign systems on an interim basis. The Ericsson system does the following:

- [ ] Call forwarding when fully automatic.
- [ ] Local automatic temporary registration when fully automatic.
- [ ] Local manual temporary registration when fully automatic.
- [ ] Automatic origination and manual call reception.
- [ ] Totally manual.
- [ ] Automatic origination with no reception.
- [ ] Manual origination with no reception.
- [ ] Roaming service denied.

The AXE 10 switch also offers the following:

- [ ] System Overview.
- [ ] Normal calls for mobiles.
- [ ] VIP call for mobiles.
- [ ] Updates traffic.
- [ ] Calls mobile subscriber.
- [ ] Automatic calls to roamers.
- [ ] Sends mobile calls.
- [ ] Switches calls in progress for handoffs.
- [ ] Administers a number analysis table in MTS.
- [ ] Handles subscriber data for home mobiles.
- [ ] Administers data for traffice areas, base stations, channels.
- [ ] Exchange data with MTS.
- [ ] Emergency states.
- [ ] Roaming data messages between MTX equipments.
- [ ] Handles visiting mobile data.
- [ ] Charging data.
- [ ] Toll ticketing.
- [ ] Output devices.
- [ ] Maintains interchange trunks.
- [ ] Disturbance supervision.
- [ ] Quality and seizure supervision.
- [ ] Testing.
- [ ] Auto transmission measurements.

Base Stations

These are the cell sites that relay radio signals between the telephone exchanges (MTX) and the mobiles, and also supervise

radios via supervisory audio tones (SAT) (Fig. 4-16). They consist of the following units:
- ☐ Transmitter.
- ☐ Receiver.
- ☐ Power amplifiers.
- ☐ Power Supply.
- ☐ Message distributor.
- ☐ Control unit.
- ☐ Transmitter combiner.
- ☐ Receiver multicoupler.
- ☐ Signal strength receiver.
- ☐ Antenna system.
- ☐ Duplex filter.
- ☐ Diversity receiver system.

Each channel is equipped with a transmitter, a receiver, and an optional diversity receiver, a control unit, and a dc/dc converter. Its message distributor interfaces between the mobiles and MTX phone exchange. The control unit generates SAT and evaluates the tone returned from any mobile. Sixteen antennas may be connected to a common antenna. The receiver and its multicoupler permits up to 17 receivers on a common antenna. Up to 16 low power (5 W) channels may be mounted within four racks together, along with the receive net, reference oscillator, and the dc/dc converters. Line inputs are 4-wire at 600 Ω impedance, and signaling on lines is at 2400 bits/sec. Total bandwidth 10 MHz for 333 channels, with duplex separation of 45 MHz.

Intermodulation products are less than 72 dB from the fundamental, adjacent channel noise less than 35 dB at 20 dB overmodulation, and FM noise is less than 50 dB. Frequency stability amounts to 0.1 ppm at temperatures between +10 and +50 degrees C. Receive adjacent channel attenuation greater than 35 dB. Low and high power consumption amounts to 40 W versus 300 W per channel.

GENERAL ELECTRIC AND NORTHERN TELECOM SYSTEM

Northern Telecom and General Electric have made a joint entry into the cellular marketing contest. Northern Telecom, with 100 years of experience in telecommunications, offers digital switches built to telephone company standards; and General Electric offers a line of cellular radio equipment based on the GE-MARC V series of trunked radio communication systems. All this, says Northern Telecom, means continuity, compatibility, congeniality,

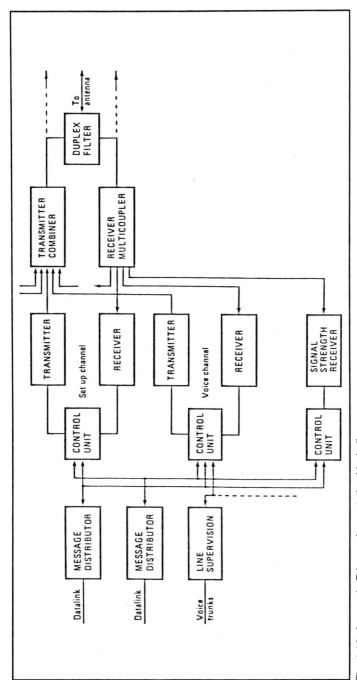

Fig. 4-16. Anaconda-Ericcson base station block diagram.

control, and cost effectiveness: the five Cs of system planning. The heart of all this, of course, is the MTX Mobile Telephone Exchange, which is actually a computer-based phone switching office that controls all connections between the telephone company and mobile subscribers, or simply mobile subscribers alone that talk to one another over extended local distances via the serving cell.

Northern Telecom shows the MTX and various cell sites grouped around it with signals flowing in and out of the associated mobiles in Fig. 4-17. The cell site rf system, shown in Fig. 4-18, has transmitter combiners, receiver multicouplers, coaxial cable, and antenna, with a cell site controller operating at all times under direction of the MTX. It may turn radio transmitters on and off, supervise calls, inject data on control and voice channels, and institute diagnostic tests on cell site equipment. Each cell site is usually connected to the MTX via leased 4-wire telephone circuits, with one 4-wire circuit required for each voice channel. If redundant controllers are used, then two 4-wire telephone circuits are needed. Each radio voice channel handles a single conversation, and each cell site has one radio transmitter and two radio receivers, the latter being tuned to the same frequency and whichever receives the stronger signal is selected as the communications point.

Alternatives to straight 4-wire arrangements are T1 digital lines equipped to handle 24 4-wire telephone circuits on a single 4-wire phone line. There are, however, channel banks needed at the cell site to change analog into digital data and special amplifiers-repeaters are placed about every mile because of the very high data bit rates. There are also private microwave systems that offer the equivalent of many 4-wire circuits on a single link, but this also requires associated multiplex equipment for baseband to 4-wire conversion. Both systems are expensive.

Modularity

In response to these traffic-handling problems, Northern Telecom has developed what it calls an Open World, Design Modularity System (DMS) the outgrowth of the DMS-100/200 and/or SL-100 telephone switches. The product comes in modular software and hardware packages so that changes and advances may evolve without major system redesign. According to Northern Telecom the DMS-100 family "combines the versatility of centralized stored-program control with distributed peripheral processing." This flexibility permits the system to be used as a "local, toll combined local/toll, military, or international gateway switch." Interbay

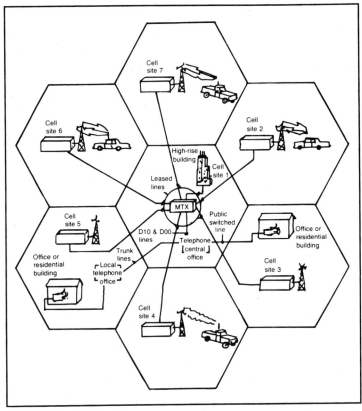

Fig. 4-17. Typical cellular system as described by Northern Telecom.

cabling allows on-site plug-ins for exchange or changed modules, and even central control comes equipped with software memory packages offering a "comprehensive range of software programs." The DMS-100 configuration is designated as a local class five system that's effective from 2,000 to 100,000 lines; while the DMS-100/200 furnishes a combined local/toll system, with size being determined by "call carrying and network limits." Maintenance and administration operations are centralized through the use of a Maintenance and Administrative Position (MAP) procedure.

A special PROTEL software high level language, developed for the DMS-100 group, supplies easily manageable software. During compile time, the operator may do extensive checking to tag errors quickly within the single entry-exit control structure. Such programs are supplied as modules which may be changed, recompiled, and loaded to handle any necessary additions and deletions.

Modules contain special program codes and data needed to execute specific phone operations or sequences. Separate but compatible modules supply system flexibility. Such features and services of the DMS-100/200/250 and SL-100 are "interchangeable with DMS-MTX modular software," thereby offering multisystem compatibility. Programs written in PROTEL are loaded into "organized modules" which may be changed or recompiled as needed. Each module contains its own information that will also blend with other individual modules, or optional modules having occasional uses or those for feature-dependent functions.

The Cell Site Controller

The cell site controller is the software-controlled interface between fixed rf frequencies and the MTX. It receives, interprets, and executes MTX instructions, including, turning on and off the voice channel rf and monitoring and reporting signal strength readings from other serving cell subscribers that may become handoffs. Based on the DMS Remote Cluster Controller previously developed by Northern Telecom, CSC may be used in simplex or duplex as the occasion requires.

CSC software measures signal strengths of subscribers assigned to the particular serving cell; reports handoff availability to the MTX; controls rf for voice and control channels; performs diagnostic; and reports any failures to the MTX.

Usually, a nonredundant CSC has only one data link going to the MTX, and MTX messages are retransmitted to any land station unit in its radio channel or reproduced as a command from the cell site controller. However, messages between the cell site controller and radio channel unit have a different protocol from messages passing between CSC and MTX. Modems operating at either 2400 or 4800 baud connect data circuits to CSC data link controllers.

Radio equipment is needed for call setup and paging, voice communications, and mobile signal strength scanning. All three channel units are identical, each is fully synthesized and can transmit or receive any of the 666 cellular FCC-authorized channels. The only difference is their mission and possibly their locations. Each cell site, of course, requires at least one control channel unit for data reception and transmission between mobiles and cell site controller. It also needs one or more voice channels which is/are also responsible for monitoring mobile signal strengths in the serving cell area. This data flows to the CSC in digital form whenever requested for when a handoff is required.

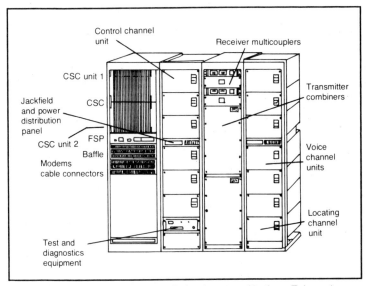

Fig. 4-18. Hardware for a normal cell site. (courtesy Northern Telecom)

A locating channel unit is available, too, to scan all active mobiles working adjacent channels by signal strength measurements. The CSC then informs the MTX, which produces a working map of all active mobiles. When handoffs are needed, map information is used to execute this requirement. As tests and diagnostics are needed, the cell site controller automatically directs any such probes and reports to the MTX. Local troubleshooting is channeled through jackpanels.

Meanwhile, as many as 16 transmitters are combined on a single coaxial cable and connected to the transmit antenna, while receiver multicouplers amplify signals from the receive antennas which may serve as many as 32 separate receivers through each multicoupler.

For maintenance and administration (position), Northern Telecom also has available a DMS-100 family MAP equipped with a visual display and keyboard, a voice module, external test facilities, slave printer, and associated furniture. The status of each system can be seen, maintenance personnel can be guided to trouble spots, hard copy printouts can be made, trunk and line testing can be done, and network administration and service order changes can be executed from one command position. In large DMS offices, more than one MAP station may be used where functional separation is either possible or desirable.

System Operation

In a sense, some of this is old hat from previous chapters; but none of them offered similar information with this type of total digital switching. Therefore, we'll repeat a little of what's gone before, but amplify and better define the digital twist, trying to make the procedure as painless as we can.

Land Party Calling. When land parties call mobiles the following takes place, according to Northern Telecom:

☐ MTX receives the call from a land party.

☐ MTX deciphers the calling code and selects the appropriate mobile—provided it isn't involved in another conversation.

☐ MTX pages the mobile over the entire cell coverage area.

☐ MTX assigns an unused voice channel and notifies the mobile.

☐ The mobile responds and ringing begins.

☐ When the called subscriber picks up the phone, MTX stops ringing and conversation begins.

Mobile Party Calling. Calls from mobiles to land parties proceed in the following manner:

☐ The calling number is entered manually into mobile memory and the Send key is depressed.

☐ The MTX receives the call through the cell site and determines its validity and subsequent routing.

☐ An unused voice channel is assigned and the mobile is notified.

☐ Audible progress tones are now heard by the mobile and the land party receives ringing.

☐ Conversation begins when the land party removes the receiver from the hook.

Mobile Party Calling Another Mobile. This, of course, takes place through one or more cell sites and is then, according to NT, also picked up by the MTX. The procedure is as follows:

☐ As before, the calling number is dialed, entered into the mobile's already established memory, and the Send key is depressed.

☐ The MTX establishes all valid conditions and looks for clear transmission lines. Invalid numbers usually bring reorder requests.

☐ The intended subscriber is paged over one or more cells served by MTX.

☐ A positive response brings a nonbusy channel assignment, and each involved mobile is instructed to tune to this channel. Busy voice channels result in intercept messages via the Control Channel, or a busy signal is routed to the initiator.

☐ When both mobiles have executed this command, the intended mobile is alerted to receive the call.

☐ When the receiver hook is lifted, conversation may begin.

Handoffs and Disconnects. During each call, the MTX monitors signal strengths on receive channels so that communications don't drop below threshold for any extended period. Should this occur, and if another channel is available, the MTX automatically institutes a *handoff*, routing the call via an alternate cell site. Such action takes place in the following manner:

☐ The involved mobile leaves the serving cell site and signal levels drop accordingly.

☐ The MTX seeks and finds a new cell to handle the call.

☐ The MTX assigns this new channel and requires the affected mobile to tune in.

☐ The previous cell then acknowledges that said mobile has left its area, and when the MTX finds that the mobile is locked into the newly-assigned voice channel, the called party is connected directly to the new channel. The previous voice channel is made available for other users.

Northern Telecom says that handoff parameters "are programmed on a per-cell site basis; thus allowing for optimization of handoffs based on cell site size, cell site traffic loading, and terrain." Also to prevent call loss or blocking during handoffs, calls involved are given priority and balanced so that continuity will be maintained. All established call signal strengths are measured periodically by the serving voice channel unit to determine if possible control changes are required.

Disconnects take place usually upon call termination and the circuits and radio channels are immediately freed for other duties. A disconnect signal also goes to the mobiles.

When a mobile subscriber goes on-hook to end a conversation the MTX terminates billing, starts idling procedures for the other party within the usual format, and places both parties in idle after the required acknowledgments.

When a land line disconnects, the MTX terminates billing and begins the mobile idling procedure resulting in channel freeing and a termination signal to the land station. The mobile acknowledges the disconnect and turns off its transmitter; the MTX places the mobile in off status.

System Features

Northern Telecom and General Electric will be offering the following expanded list of features for their system, some of which appear to be unique over competitive systems. A general listing follows:

> Call forwarding, call transfer, 3-way calling, call waiting, abbreviated dialing, mobile limited-duration discount service, roamer handing and registration, call forwarding if no answer, mobile operator position services for credit card calling, third party billing, malicious mobile call tracing, data communications, voice storage and retrieval, load balancing, dynamic power control, handoffs, least cost routing, time of day routing, billing, disk search of billing records, automatic trunk testing, pending order file, journal file, operational measurements.

Warranties are 12 months parts and labor. Technical assistance is available, and annual contracts may be procured from Northern Telecom and G.E., or repair and return programs can be arranged if a customer maintains his own equipment.

An informative graph of what some of this is all about appears in Fig. 4-19. The graph shows subscribers per voice channel versus the number of voice channels at a random cell site. For up to 15 subscribers 5 voice channels, the curve is relatively linear. But, thereafter, efficiency begins to drop markedly until the curve begins to level off at 25 subscribers and 25 channels. However, multiplying 25 × 25 produces 625 total subscriber operators for this particular omni cell, with an Erlang loading for each subscriber of 0.026 and a very small probability of blocking.

Central Processing Unit. The heart of the DMS-MTX system is the CPU. It controls all other parts of the system on the basis of instructions received from the program store memory module (PS). Actually there are two of these CPUs operating as a Central Control Complex (CCC) in matched mode, each with solid state dedicated memories. In this CCC complex there is also a data store memory module (DS), and a central message controller (CMC).

Synchronized CPUs have PS and DS memories, and additional memory modules may be added as the system grows. Each CPU (only one is fully operational at a time) is a high speed, 16-bit stacked, microprogrammable processor having two parallel ports. One port is designed for autonomous receipt of instructions, while the other accesses data and in/output devices. However, because

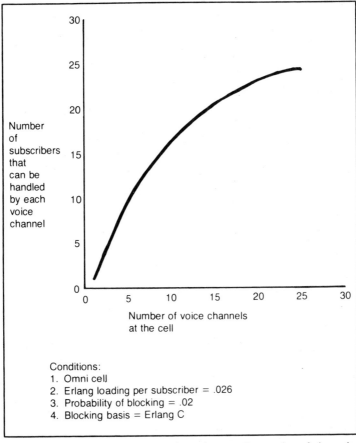

Fig. 4-19. Graph of subscribers per channel versus the number of channels. (courtesy Northern Telecom)

they are a matched pair and running online in sync, one CPU is continuously compared with the other over matched circuits. Improper operation such as an out-of-sync condition immediately commences remedial action. Only the active CPU may communicate with the processor memories. At this point both data stores contain the same information, the two CPUs are connected by a mate exchange bus, and the two units are synchronized for data exchange and maintenance. Data ports compare information and parity between CPUs and, if there is a mismatch, an interrupt occurs and recovery instituted. In this way the bad CPU will be identified, the system reconfigured, and all processing ability recovered. This self check tries to identify hardware as well as software problems, but both CPUs must be operational for this to occur. In simplex operation, the backup CPU is out of service and offline, and system protective features are unavailable.

Every CPU has program store memory (PS) which contains program instructions for call processing, administration, maintenance, and system operation. Channels are selected and ringing commenced in call processing; administration tells the CPU how to access data store memory (DS) for updates or data changes; and maintenance includes fault diagnosis following problem location.

The data store memory accumulates call-type information in addition to customer data and office parameters. It remembers such items as the originating trunk, the called number, the terminating trunk, channel numbers, and the beginning and ending of each call. This transient information, of course, is removed from the DS whenever a call is completed. Customer data includes class of service, phone numbers, trunk identification, and terminals to which the CPU is connected. DS and PS "shelves" may each contain up to 4 million words of memory.

Peripheral Modules. The peripheral modules (PM), contain digital trunk controllers (DTC), trunk modules (TM), and maintenance trunk modules (MTM). They interface among switching networks, digital carriers and analog trunks. Control microprocessors are responsible for state changes in scanning trunks, timing functions during call processing, digital tones, signaling, and network integrity checks.

DTC has two functions: it interfaces up to 16 network ports of 30 (plus 2) channels, or 480 speech channels; and it supports up to 20 ports of 24 channels each for connecting digital trunk lines directly to remote units.

TM and MTM modules house service electronics such as receivers, audio tone detectors, announcing trunks and test circuits. Each may interface 30 service circuits with one 32-channel 2.56 Mbits speech link to the overall network. Maximum numbers of five TMs or four MTMs may be placed in each equipment rack bay.

Software. A high level PROTEL language has been specifically developed for the DMS-100 family of switching systems, principally for structured programming techniques in entry/exit control and to identify errors at compile time. Each PROTEL program is organized into a module, and may be changed, recompiled, and loaded as needed. Each module contains program codes and data for telephone functions. Base area operating software permits different hardware units to communicate with each other upon command. If offers data base manipulation, software updating, hardware communications protocol, MAP command interpretation, dynamic storage assignments, plus backup and maintenance.

Switching software primarily occurs in the common area which operates as the call processing base for switching services and the action coordinator for PM (peripheral module) functions. A block diagram illustrating much of this description is shown in Fig. 4-20. In it you see I/O, Call processing, billing, maintenance, MAP, file system, operational measurements, and data control, with temporary data processed on the right.

MAP Diagnosis. MAP stands for maintenance and administration. It has many built-in "administration, hardware, maintenance, and software features designed to minimize operating costs." Maintenance operations (functions) are coded in modules corresponding to system hardware functions. They are subdivided

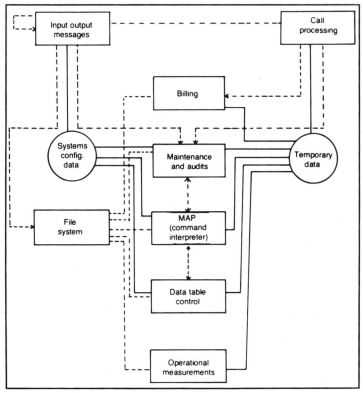

Fig. 4-20. DMS-MTX software overview. (courtesy Northern Telecom)

into subsystems such as routine testing, fault detection, analysis, effort reporting, diagnostic reporting, and/or system status reporting. Maintenance people have diagnostic information at such a level that problems may be corrected by simple replacement of an electronic printed circuit card. Current system status can be shown on a visual display unit and various information may be accessed from input commands to the VDU.

The MAP, itself, has four basic components: the VDU, a voice communications module, test facilities, and position furniture. A CRT offers visual display, a detachable keyboard provides information entry, and a printer (in parallel with the VDU) produces hard copy for permanent records. Initially, relevant information is displayed on the cathode ray tube. A menu then suggests specific interrogations, and the operator may then zoom in at lower subsystem levels to find specific faults. The voice communications module offers hand-off and handset operation as well as auto dialing. Three double jacks are mounted alongside the table surface for direct links between test trunks and portable testers. The test furniture consists of some 20 square feet of table surface, documentation storage, and an adjustable drop surface for the VDU keyboard. Various interfaces such as EIA RS-232, loop current, and data set are used, depending on the distances involved. For unattended DMS-MTX equipment, several of them may be monitored and controlled by MAPs at some central maintenance center or depot. For standard routines, normal fault detection, analysis, and routes may be executed periodically by system demand, or on request. Emergency routines are also available should a MAP itself fail. There are also standard visual and audible alarms included for apparent troubles and thresholds may be set by customer option. Transfer or cancellation of visual and audible alarms may be key-activated at the MAP or via an alarm control or display unit.

Internal testing results in system-initiated call tracing tests, fault isolation, and recovery checks of vital subsystems. External testing includes industry standard transmissions, noise, and signaling tests using 100 series test lines and remote module tests.

MAP instructions, as an example, permit the following: transmission tests of voice channels; HDLC link tests; redundant data link switching; redundant CSC switching; actual CSC tests; and hardware and software rf base station equipment tests.

Stationary and Mobile Radios

Part and parcel of what Northern Telecom calls its "Cellular

Radiotelephone System" is General Electric's UHF radio equipment at each cell site and the mobile UHF radios in each serving area. Based on extensive experience in 800 MHz radiotelephone products, G.E. will supply one transmitter and two receivers for each assigned voice channel and one or more transceivers for paging/access channels. Individual reference oscillators and frequency synthesizers are used throughout.

Features. Mobile telephones offer such features as least cost routing, call recording, flex dialing, self diagnostics and remote maintenance. Stations deliver call forwarding, conference calling, waiting, time reminders, message storing/retrieving, roaming, restricted access codes, and speed calls. The mobiles also have paged area code display, single volume control with memory, data/voice recorder connection for computer access interface, DTMF end-to-end signaling access, electronic exclusion lock, auto horn alert, and an electronic clock that shows the time of day, call duration and time accumulation. Figure 4-21 shows a G.E. mobile transceiver.

Specifications. Mobile specifications include -116 dBm for 12 dB SINAD sensitivity, talk mode drain of 3 amperes, off mode drain of 15 mA, audio response of 6 dB/octave, and RSSI range of 110 dBm to -30 dBm for the receiver, 3 W power output for the transmitter, maximum distortion of 5 percent, modulation deviation of $\pm$ 12 kHz peak, and FM hum and noise of -32 dB.

The cell site transceiver units have ± 1.5 percent frequency stability, a power output of 4.5 to 45 watts (adjustable), -40 dB FM hum and noise, and a transmitter carrier shift of 500 Hz (maximum). The other specifications are generally comparable to those of the mobiles, where applicable.

Signal Flow. To give you a little better idea of what the hand-held personals and mobiles look like, I importuned G.E. for a block diagram showing at least signal flow. Figure 4-22 is the answer to my request. The description of its contents that follows is strictly mine, however.

Beginning from the left, antenna inputs and outputs flow through the duplexer and either from the TX power amplifier or into a sensitive rf receive stage. Incoming signals are mixed with variously crystal-controlled outputs from the 666-channel frequency synthesizer, mixed for downconversion and put through a spur and harmonic-suppressing bandpass filter, and on into the intermediate frequency logarithmic amplifier. Here downconverted audio or logic is amplified and passed through to the audio detector which, apparently, has three functions: audio can be routed to the audio

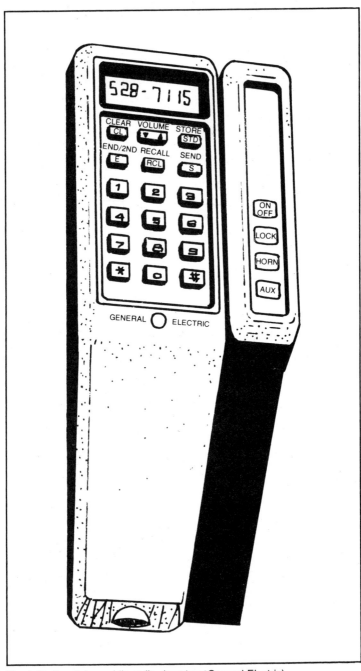

Fig. 4-21. A cellular mobile radio. (courtesy General Electric)

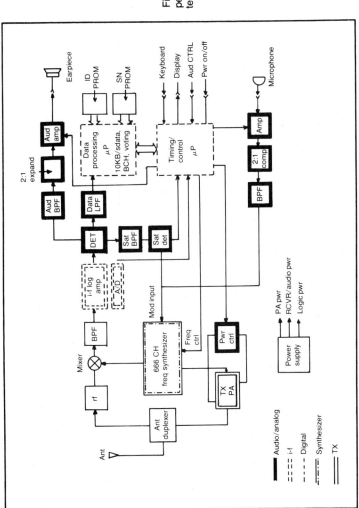

Fig. 4-22. Typical block diagram of a personal/mobile transceiver (courtesy General Electric).

bandpass filter, doubly expanded and final-amplified for the equipment speaker; data can proceed through a low pass filter and on into the 10-kilobit data processing microprocessor; and a portion can be taken off as a 5.970, 6, or 6.030-kHz SAT tone, which is transmitted back to the cell site to close the communications loop and also proceeds to the timing control. A 10 kHz signaling tone originated by the mobile in response to an incoming call alert is also involved. Audio/data can now be fully processed as the data microprocessor also recognizes identification symbols and checks signal-to-noise, while interacting with timing/control.

Timing/control supplies outputs to the transmit amplifier, power control, and frequency synthesizer, in addition to receiving audio, power on/off, and keyboard operations from other transceiver functions. Speech is 2:1 compressed, bandpass filtered and delivered to the frequency synthesizer, where it is modulated on a specified channel frequency and then power-amplified by the transmitter power amplifier for the antenna duplexer and the antenna itself. Conversation or data flow now may flow freely until caller or called goes on-hook and terminates the exchange.

Trunked Mobile Radio

General Electric is also very active in the trunked 800 MHz system which is a purely business radio exchange type, but differs markedly from cellular in that there are no handoffs or cell sites, just repeaters with fixed channel assignments. Several users may access or be assigned to an individual channel, making others wait. G.E., however, has developed a system which may include from 5 to 20 repeaters, permitting users access to any free channel available. A total of 200 channels have been FCC-allocated to this service since 1976. G.E. says "the low cost, expandable modular design can handle up to 256 phone lines, one per customer or customer group as the FCC rule dictates. A photo of a G.E. trunked radio and vehicular phone in one unit appears in Fig. 4-23.

ITT'S CELLTREX

Called CELLTREX by parent ITT, and identified as its 1210 digital switch, cellular mobile radio service systems now have another U.S. designed and manufactured fully automatic radiotelephone switching and control system for use in cellular mobile radio systems. Designed to serve up to 100,000 subscribers, this frequency distribution setup contains three key elements: the NCS

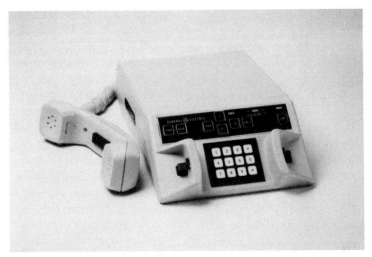

Fig. 4-23. Trunked radio and vehicular phone (courtesy General Electric)

(network control switch); the RSG (remote switch groups) and the CSC (cell site controllers).

Fully digital with generic software, the network control switch furnishes management and control services for remote switch groups, cell sites and local traffic, in addition to central maintenance, administration, traffic/billing operations, and system control for hand-off mobile unit locations and outside network access.

Instead of being returned to some network control switch for routing and wireline network entry, this network access system requires back-haul voice circuits and limits the entire system correctly to contiguous (boundary) area coverage. Further, when RSG's are used, localized calls do not require back-haul voice circuits, and control links from the network control switch supply remote switch group connections. RSG, of course, eliminates the back-haul voice circuit concept, allowing out-of-the-way communities to use local mobiles at the local office, including local directory listing. The RSG furnishes local switching, message relaying, and handoff switching.

Microprocessors also control various cell sites by opening entry points for radio-telephone communications. They report to an NCS or RSG, as required. Tasks such as data collection location, message relaying and reformat, test/maintenance and signal fades are also undertaken by such controllers. Then, by using the three NCS, RSG, and CSC elements, the ITT 1210 offers extensive network control, even permitting cellular radio in areas usually too small to permit conventional cellular service. And with CELL-TREX software used in a conventional Class 5 or 4/5 1210/64 digital exchange, startup costs are considerably reduced, according to the company, and cellular service is possible in such normally excluded areas.

ITT's NCS has redundant system control, remote capability, master/slave sync/ nonblocking digital matrix, shared services, module sizing, adaptable network access, handoff sync, mobile unit location data analysis, LAMA/CAMA, local/polling, custom calling, remote alarms, full peripheral support, centralized maintenance/administration, optional assistance operator, and roamers.

The remote switch group (RSG) offers community switching, intra-network traffic, multiple network access, cost effectiveness, full NCS growth, if required, and handoff switching between RSGs.

The CSC cell site controller permits base radio remotes, compatible rf interface, analog operation to base radios, T-1 to RSG or NCS, multiple cell sites per CSC, and microprocessor control.

E.F. JOHNSON EQUIPMENT

With E.F. Johnson now joined by merger to a newly organized subsidiary of Western Union Corporation through a 1:1 conversion of common stock shares, an outstanding 2-way communications radio manufacturer has been joined with a huge communications common carrier. And since Western Union Telegraph Co., a subsidiary, provides telecommunications systems and services to business, government, and the public, a major marketing effort of traffic and hardware is expected.

I would like to offer a complete rundown on Johnson's mobile and base station equipment for cellular, but the Company apparently isn't ready to talk about a broad range of products—at least at this time. Therefore, the best we can do is quote specifications from mobile and base station handouts which may be of some interest.

Performance specifications of the Model 1154 cellular mobile telephone are as follows:

Transceiver: 11.4" L × 10.6" W × 3.1" H. Control Unit (dial-in-base); 8.3" L × 4.3" W × 3.3" H. Control Unit (dial-in-headset): 8.8" L × 2.8" W × 1.5" H. Operating temperatures −35 to +60°C; and power requirements 9 V to 16.5 V at 3 amperes maximum.

Transmitter power out is 3 W into 50 ohms; spurs & harmonics −65 dB minimum; audio response ± 1 dB of + 6 dB/octave.

Receiver measured at 12 dB SINAD, −116 dBm sensitivity;

spurs and image rejection, 100 dB minimum; audio response, same as transmitter.

Information for cell site radios and their amplifiers was considerably more specific. Specifications, however, are subject to change since neither these units nor the Johnson mobile have been type accepted by the FCC yet. You will note that there is little current demand even in transmit for the transceiver, but do remember that a power amplifier must be attached for this base station to function normally. So 8.6 amperes in full duplex is more like the actual power drain rather than 0.7 A or 0.9 A, whichever you see first.

The Model 1162 transceiver measures 15.8" H × 1.8" W × 14" D, and weighs 45 lbs. At 24 Vdc, the transmitter requires 0.9 A, the receiver, 0.7 A, and full duplex, 1.6 A. Temperature range extends from −30°C to +60°C, and specified frequency range extends in transmit from 870 to 890 MHz, and in receive from 825 MHz to 845 MHz.

Receiver sensitivity (12 dB SINAD), −116 dBm; selectivity at ± 60 kHz, −65 dB; intermod, −65 dB; audio distortion 2.5 percent; and audio response between 200 kHz and 20 kHz, ± 2 dB.

Transmitter rf out, 1.5W; maximum deviation ± 12 kHz; frequency stability + 1.5 ppm.

The *power amplifier* measures 3.9" H × 5.1" W × 12.4" D, and weighs 7.5 lbs. In transmit it requires 7 amperes, and the same in duplex at 24 volts. Temperature range is the same as the transceiver, and it covers the frequency range in transmit from 870-890 MHz. Transmit output power ranges between 7 to 45 watts.

FUJITSU TEN'S AVM

While not strickly cellular, the Japanese are using an interesting method of automatic vehicle monitoring and dispatch arrangement which could, conceivably, become part of a cellular-type system under favorable operating conditions. Whether this could or would occur in the U.S. or not is purely speculative, but the concept is worth reporting, nonetheless, and should be considered as a worthy part of our 2-way communications explosion of the '80s.

Known as the "TEN" AVM system, and developed by Fujitsu Ten, taxis may be dispatched efficiently to any urban area covered by special radio even during rush hours, offering expeditious service to riders and allowing each cab to proceed to its new destination in order of rotation with considerable savings in travel and waiting time.

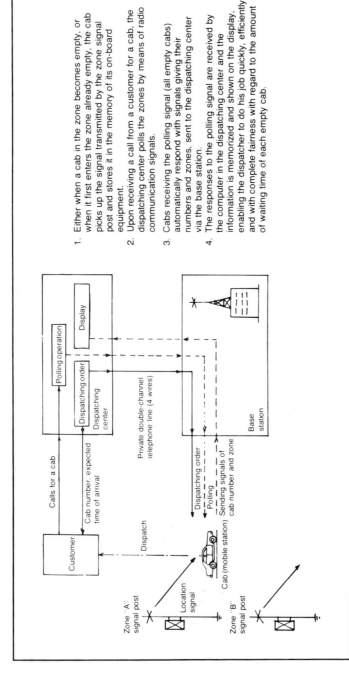

Fig. 4-24. AVM system working diagram.

1. Either when a cab in the zone becomes empty, or when it first enters the zone already empty, the cab picks up the signal transmitted by the zone signal post and stores it in the memory of its on-board equipment.
2. Upon receiving a call from a customer for a cab, the dispatching center polls the zones by means of radio communication signals.
3. Cabs receiving the polling signal (all empty cabs) automatically respond with signals giving their numbers and zones, sent to the dispatching center via the base station.
4. The responses to the polling signal are received by the computer in the dispatching center and the information is memorized and shown on the display, enabling the dispatcher to do his job quickly, efficiently and with complete fairness with regard to the amount of waiting time of each empty cab.

Computer controlled, the system tracks each taxi in its assigned zone, and when in or out of service can communicate with that particular vehicle, delivering destination instructions, noting any emergencies, and monitoring all movements.

When the operator/dispatcher receives a call, names, addresses and number of vehicles are immediately noted, the zone is determined via keyboard or light pen and the "up" cab among the top five in any zone is selected to answer the call. The cab number and estimated time of arrival is reported to the caller, and next the cab then goes to the head of the list, receiving the next available call in its zone.

The procedure is illustrated somewhat more graphically in Fig. 4-24. Here you see Zones A and B representing signal locations of two groups of taxis. The polling operation determines the location and availability of each taxi, and the cab number and zone are recorded on a display in the dispatching center. According to priority, one taxi is selected, a dispatching order transmitted and the customer is notified immediately as to cab identity and its estimated time of arrival.

Dispatching centers contain control equipment, telephone line terminals, a computer processing unit, a visual display, a console typewriter, and a control table. The taxi has its zone signal receiver, a signal processor, two-way radio, and its operator, while each zone signaling post has a transmitter and antenna. Visual displays at the dispatch centers are in three colors, with cabs identified by numbers and waiting times. As many as seven zones may be shown and a console typewriter furnishes print outs of dispatch operations and emergency records. When there are emergencies, both the cab number and zone are registered on the display in flashing red letters.

The Japanese say this dispatching method represents an expeditious and very fair way of using all available fleet service to take care of a large number of riders wherever they may be in an urban area. With cellular added to regular 2-way radio, the system could easily handle suburban traffic also. But for this we'll just have to wait and see.

Chapter 5

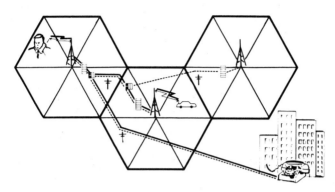

General Electric's CB-Telephone Proposal

Dubbed Cellular Junior by some, but really a low-cost, high frequency public mobile radio with phone connections, General Electric has introduced a Personal Radio Communications Service (PRCS) that the Federal Communications Commission has accepted for Notice of Proposed Rulemaking. Taken on January 20, 1983, the FCC action permits the application to be commented upon by all interested parties, and the Regulatory Agency, itself, will probably make its final decision of approval or disapproval sometime in 1984.

In its original proposal, G.E. requested two 4.5 MHz set-asides separated by 45 MHz in the 900 MHz spectrum. But the FCC apparently looks favorably on only 8 MHz for the service, and 133 instead of 149 channels. Nevertheless, only two of the regular seven commissioners are so far expressing any real reservations and PRCS seems to have a better than 50-50 chance. Other would-be 900 MHz CB manufacturers are expected to also request approval for their competing systems provided, of course, the omens are favorable.

Then there is a possibility that dissenting commissioners may acquire other allies, and the entire spectrum around 900 MHz gain approval for government/non-government fixed service use and/or permit public air-ground telephones such as those proposed by ARINC or its competitor Airfone, Inc. FCC bureaus are now working on a special study of spectrum allocations, and after this is completed and all comments are in, then a decision will be forthcoming on who gets what—probably after January or February of 1984.

THE USER MARKET

There appears to be interest for the PRCS among potential users. The result of a survey point to who those potential users are. The price of this equipment may also be a very important factor to consider.

Opinion Research Survey

In March of 1982, G.E. received an Opinion Research Survey which it had previously commissioned covering a national sample of 2,509 adults who regularly operated some type of motor vehicle owned by the household. Briefly, the survey found that: 95 percent of the auto owners drive for various reasons other than going to work; almost 75 percent of these arrive at work in a household-owned auto; and some 25 percent of this group use their vehicles for additional work purposes other than commuting. As might be expected, men are by far the larger work-related commuter group, with women using vehicles more for personal reasons. The largest users were found to be in the age group between 25 and 49, with professionals topping the lists. Most commuters drive between 10 and 30 minutes with an average driving time of 20.3 minutes at a mean round trip distance of 20.9 miles.

Traffic tieups were the greatest hindrances followed by accidents and mechanical failures. When traveling, drivers are said to prefer pay station telephones over other means to reach those with whom they wish to speak. The survey showed that 64 percent use telephones and only 7 percent CB radios. Some pagers and beepers are in use, but only a small percentage compared with telephones.

However, to give you an idea of potential users, only 19 percent of those surveyed were interested in actually owning and operating a PRCS mobile or base station. Considering, however, that there are probably over 65 million households owning trucks or automobiles, even 10 percent of the total would offer a potential of 6.5 million subscribers for starters, and that could be worth a few nickels for the right entrepreneurs.

If you were a frequent home or business caller working some metropolitan area or just around the farm, wouldn't this low cost service appeal to you? Even far afield there'd be no need to stop for a pay station, just pick up your PRCS mobile and dial straight through.

The time saved plus your convenience factor would be worth repeater billing for an entire month for just one call.

One additional set of survey statistics for the record; possible purchasers were recorded as 58 percent to 41 percent male versus female in the 25 to 49 year age group, and prime buyers are in the white collar and skilled blue collar categories, followed by housewives or the retired and professionals. That should give you a pretty good idea of what PRCS could be all about. Obviously, our Knights of the Road would hardly participate, retaining major dependence on CB which they now monopolize as their 18-wheelers pound the highways.

Price

Price, in PRCS, is very likely to be the largely governing factor, just as it has become in regular CB. As an example, how many citizens band radios do you see today that are single sideband equipped, and how many persons do you hear using that superior service? A few? None?! Today most 27 MHz CB radios sell for $50 to $75 and are restricted to 40-channel AM. Generally, that's about all the general public knows, even though both distance and noise factors are infinitely superior in AM single sideband over standard carrier and dual sideband transmissions. At least, the new proposed service should appear at about 900 MHz and emit a constant carrier with audio deviation within set limits.

PSTN INTERCONNECT

This is PRCS' outstanding feature and precisely what the 900 MHz service is all about . . . Automatic interconnect between the system's base station and the public switched telephone network (PSTN). Without such an outstanding attraction, the Personal Radio Communications Service would be just another FM radio/CB affair and there would be little hope for final FCC approval at any time.

But with this addition, such interconnections may take place automatically with almost total surety that access to telephone lines can only come from a properly authorized base or mobile which has the necessary coding. Talk channels also, once assigned, are entirely private. The base station initially must determine if the telephone is on- or off-hook. If no phone call is requested, channel verification immediately occurs; but should a phone call be desired, the public switched network interface is activated to determine an available line. With an open line, code validation then certifies a proper call. If a line is not open or the call is not validated, the base station enters the channel-scanning/NACK Process to see if the station is responding to either direct or repeater talk channel requests. The 4-bit Destination Address Data (DAD) word sets to 0000 and channel code to null channel code. Here's how it all works:

☐ Mobile-to-Base Interconnects take place when the mobile offers all the correct flags and codes on some designated talk channel. The base station switches to its phone connection and passes a dial tone back to the mobile. If there is no busy signal, busy-out action, or call-waiting tone, the mobile delays 750 milliseconds to be sure of the tone and then transmits the full digital phone number of the party called.

☐ Base-to-Mobile Interconnects begin with calls coming from PSTN to the base station for transfer to a mobile. The base station accepts the call, signals the caller that transfer action has begun, and then transmits a normal rf carrier complete with the control message to one (or more) programmed mobile stations(s) authorized as this base station's particular receivers. A number of such mobiles may be included or excluded, depending on programs already stored in the base station's memory. But callers may not individually select specific mobiles with the same phone numbers.

☐ Local or Repeater calls are handled differently, however. A local call in full duplex has both mobile and base stations remaining on during the entire time any conversation is taking place. When terminated, the mobile transmits a disconnect to base, which then terminates the telephone signal by going on-hook.

On a repeater channel, the base interrupts its transmissions about every half second and turns on the receiver for approximately 10 milliseconds to listen for mobile station transmissions. Any mobile transmitter tone will turn off the base transmitter for as long as the mobile is calling. Once the mobile quits, the base can begin transmitting again with the usual intermittent interruptions. Such action is necessary to save spectrum since using a repeater would ordinarily require two channel pairs, or four individual frequencies.

SYSTEM OPERATION

In G.E.'s original application to the FCC, Docket 79-140, dated May 11, 1982, it requested a Personal Radio Communications Service consisting of 149 duplex channels for selective calling that could access two simplex channels for direct communications. Of the 149, 5 would be control channels, 32 used for direct calling between stations, and 112 for access to other stations through repeaters. Mobile and base stations combined would be expected to sell for approximately $400 and differ only in ac versus dc power supplies and a telephone hookup. Figure 5-1 pictures a transceiver and a handset.

In either the 133- or the 149-channel versions, each channel would have a 30 kHz spacing and FM speech deviation to 8 kHz, with transceiver (base/mobile) accuracies pegged at 5 ppm (0.0005 percent) and repeaters at 1.5 ppm, or 0.00015 percent. Regular FM test sets should handle this equipment except for the digital identification and signaling which will be discussed as the chapter progresses.

When there are busy channels blocking normal communications, the system will seek reusable channels automatically; base and mobile stations are operational upon purchase from any source whatsoever (plus the simple installation of a small antenna); and repeater fees are expected to amount to only some $10 per month. In operation, mobile to base station communications should be good for from 2 to 5 miles and via repeaters, at least 15 miles. Requested power for the mobile/base combination is 10 watts each, and repeaters may offer as much as 100 watts, depending on both the service area intended and surrounding terrain. Besides FM, other systems considered were single sideband (very expensive at 900 MHz and extremely high frequency accuracy required), spread spectrum (too many users could jam all channels), and digital, which was only evaluated for informational coding (not modulation) and which may be either phase or frequency shiftkeying.

Fig. 5-1. A photo of General Electric's PRCS equipment. (courtesy G.E.)

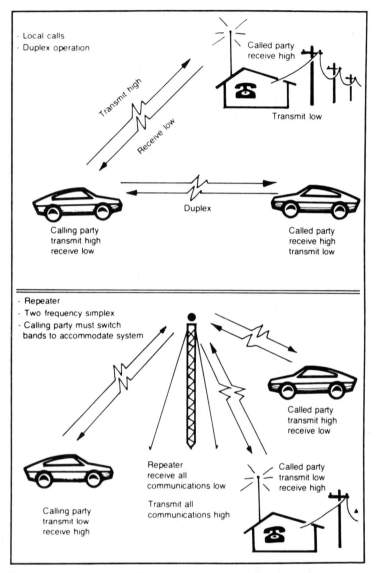

Fig. 5-2. Illustrative sketch of how the system operates. (courtesy G.E.)

In placing a call from one PRCS station to another, as Fig. 5-2, the transceiver's selector is turned to trunked channels and the address of the local station entered on the keypad. If connections to the telephone through the user's base station are required, an additional code number must also be entered. Thereafter, the addressed and transmitting stations will establish proper identification, choose a single channel for conversation, and/or dial a phone number for the address through the transmitter's base station and wait until conversation begins. Unless such communications occur at peak use time, such 2-way talk continues until the end of normal discourse. In peak periods, however, conversations through the repeater station are automatically limited to three minutes to give the other fellow a chance and avoid jamming the entire system and keeping other users arbitrarily off the air. In other words, the repeater senses extraordinary channel use and responds by curtailing talk time.

In a study conducted for General Electric, PRCS users are expected to place between one and three calls each peak hour (peak hour defined as 6-9 A.M. and 3-6 P.M. weekdays) and approximately 2-6 calls per day. If, however, subscribers placed only 1.2 calls-per-station (CPS) during peak hours with an average duration of 80 seconds, G.E. says that 9 percent penetration could be reached "in all but the top four markets before quality of service would drop below 80 percent."

These statistics should offer some idea of what the system can do and its evident limitations. On the other hand, relatively few commuters (on a comparative basis) would normally use this service, and those who wanted both more distance and freedom from interruption would automatically use either General Mobile or Cellular anyway. So this specialized, low cost service is specifically designed to do what its proponents suggest—serve commuters and small business people in metropolitan areas (generally) at the lowest cost.

Licensing Only for Repeaters

Licensing for the proposed service is presently considered only for repeaters and not for the base or the mobiles. Like Citizens Band Radio, ordinary users won't be required to notify Gettysburg, Pa., of their PRCS units and may go on the air without audible call letters. There will, however, be a 3-digit transmitted identification number assigned by the manufacturer to each radio called TIN. The purchaser/user would then add his own individual 8-digit number, usually his telephone number plus one digit. This 11-digit identifier would or could automatically become the station license identification for the FCC, the phone company, and anyone else, serving as ID, calling number, and billing address.

Since a list of such numbers and their subscriber-owners will be kept by repeater stations, repeaters may become an effective FCC and local monitor to keep the system in manageable operation. This self-policing proposal should appeal to most FCC Commissioners of the moment who seem to want deregulation of just about everything—a political course which may well return to haunt both Government and Congress as the vast network of communications in the sky and on earth continue their explosive and somewhat unsupervised growth.

Standards

Whether the FCC wants to go along or not, standards will have to be either defined or accepted as digital protocol for communications handshaking by the control channels. Every station must also have a power cutback capability so that each will emit only enough rf to reach the intended channel or fellow transceiver. Repeaters should be able to select individual channels to void transmissions by nonsubscribers and those intended for other repeaters at various localities. General Electric points out that if there is no channel gating, a single message could be transmitted or retransmitted through two repeaters, resulting in considerable interference for everyone.

Nominal power outputs for both mobile/base transceivers should also be assigned—they're now suggested as 10 watts. Even the repeaters will have to be told if the proposed limits of between 30 watts and 100 watts are suitable. Base station and telephone interconnects probably also need some specification, and patent regulations may have to be established, probably on a royalty basis, so that all participating manufacturers in the system will have an opportunity to produce a relatively uniform product. G.E. has already said it is willing to license "on reasonable terms" patents granted on the basic PRCS system. Others entering the market should probably be required to do likewise. I strongly suspect, nonetheless, that PRCS and cellular transceivers will have even less in common when uniform characteristics are selectively compared.

Automatic Transmitter Identification System

The automatic transmitter identification system (ARTIS), is another constraint the FCC must impose on the proposed service for both identification and subscriber billing/validation. This is a very effective and inexpensive means of guarding against pirates appearing on PRCS and would negate many of the problems which have occurred on the now totally unregulated CB channels. Repeaters, destined to develop as the ID mainstay, are expected to become "a highly competitive entrepreneural activity," according to G.E., and their operation can be compared to existing Specialized Mobile Relay (SMR) systems in the 800 MHz band. G.E. suggests this may become a very profitable enterprise when PRCS is fully operational and even groups with common interests could place such repeaters in operation on a share, stock, or some other agreeable financial basis.

Loosely estimated costs for such stations might amount to $1 million each depending, of course, on wattage output, operator-related or automatic equipment, modest or extensive services, etc. A few thousand subscribers at $10 to $20 a head per month should bring in a tidy sum. The FCC, by the way, has federal jurisdiction over SMRs, and G.E. suggests the same action for the PRCS equivalents. Repeaters and mobile/base stations, of course, would all have the same operational frequencies assigned to others throughout the U.S. engaged in PRCS communications.

Signaling

Although we've already given a brief description of how the system is designed to work, there are more specifics available that might be of interest to those who expect to actively participate. Therefore, all will be included somewhat succinctly both for the record and to satiate better than average curiosity. Again, all this information is supplied directly by G.E. since no one else has yet filed an application with the FCC.

The description will be based on the original application calling for two nondirected simplex channels for direct communications between mobiles or base stations tuned to identical channels and 149 duplex channels for selective-calling radio communications. Of these, 5 are control channels, 32 are local, direct calling channels, and 112 are repeater channels for extended distances. Selective channels offer access to dialed and/or trunked channels via both mobiles and base stations, in addition to repeaters. Channel selection by handshaking is to be done on the five control channels by digital signals and nowhere else. Error checking of control instructions will be accomplished between the sending and receiving stations, which will also assign the channel for communications. When local channels are not available, repeaters are contacted and a subscriber identified. The control message will then be regenerated and transmitted correctly to the receiving station.

Local Channels. The 32 local talk channels operate on standby and are not routed through repeaters. In full duplex, they remain ready for discourse between two stations whether base or mobile and are limited to the usual 2 to 5 mile ranges specified for PRCS under normal operating conditions. Within local areas, they should be reusable a number of times because of their limited range. Remember that 10 watts is all that base/mobile transceivers are permitted under present considerations, which probably won't change. Nor will there be any charge for two-way talk since messages don't go through repeaters.

Repeater Channels. These 112 channel pairs are assigned strictly for dual frequency simplex operation and do pass through repeaters, since that's what they're designed for. Such channels may only be accessed through a repeater on a peak-time restricted schedule (otherwise open) so that one subscriber may call another over distances greater than ordinary equipment-to-equipment talk and have a specific channel assigned as soon as the caller's code is recognized and a response forthcoming from the called. Ranges up to some 15 miles are probable under routine conditions, perhaps more, depending on repeater power (between 30 and 100 watts) and antenna heights which will vary with terrain and the repeater's pocketbook.

Party Lines. The two channels which are nondirected and simplex are those used for direct contact between mobiles and base stations and called Party Lines. No phone number, channel designation, or restrictions apply here, and they operate just like conventional CB on the open road. They are simply another way of talking with a neighbor or another motorist in the near vicinity. As G.E. puts it, "an adjunct to the primary selective-calling features of PRCS."

Protocol. We can now begin to talk specifically about how calls are initiated and completed with their control formats and execution. To begin a call, the calling station will transmit a control digital logic bit stream to some other station on a certain control channel specified by the final three digits in the other station's address. Local talk channels, of course, operate full duplex and operate on two frequencies: a high band and a low band, separated by 45 MHz. On repeater channels, two frequencies are also used, but operation here is simplex. Therefore, the calling transceiver will have to band switch between the two frequencies in these two modes when it is in conversation. Specific frequencies are determined by whichever mode is used: local or repeater.

Except in actual conversation, all stations except the repeater(s) monitor high band control frequencies. The repeater, of course, looks at the low band controls, and retransmits any such command on an appropriate high band. In the case of direct communications, mobiles or base stations try to find an open local talk channel before calling the repeater and then band switching.

Message Formats require the following: 96 bits of preamble to synchronize both transceivers; 32 bits of start in the form of eight 4-bit characters offering compatible operation among all transceivers in use; 44 bits of source address, including address of the transmitting station and the transmitter identification number (TIN); a 32-bit destination address to the receiving station; 12 bits to identify the talk channel suggested by the transmitter; 8 bits in the flag field for F0, F1, and F2-F7 which stand for direct or repeater contact, telephone or nontelephone interconnect at home base; a final six bits for secure phone interconnect; and the CRC-16 which are 16 redundancy check bits to detect any errors in the message.

Every control message has two separate segments: the first contains both preamble and start, to be followed by the second with the balance of the 112-bit message generated three successive times, making a total of 464 bits transmitted in each control message. The receiving station synchronizes on the preamble, then checks for a unique start code appearing in the start field, and the final 112 bit positions are compared with bits in other message formats in an attempt to arrive at a correct message. Finally, messages are re-checked by CRC-16 error detection redundancy bits. Should two stations in the same general area transmit simultaneously, both transmissions will cancel out, and there will be no acknowledgment. On second or third trials, however, their messages are randomly delayed for periods up to two seconds so one or both will get through to a selected receiver. If there is no answer after four attempts, the transmitter will try and find a repeater channel. If one isn't available, the call will busy out for 20 seconds before it can be attempted once again.

PRCS Calling Rules for calling and receiving stations must be followed precisely for satisfactory contact. A control channel is selected by the final three digits in the receiver's address which should always be monitoring some chosen control channel. Message format, source and destination must be deciphered, flag field set for contact, and talk channel assigned. Local talk channels are switched to high bands for transmit and low bands for receive. Repeater talk channel users transmit on low bands and receive on high bands into the repeater and the repeater retransmits on high bands. In calling, the receiver's address is sent, then the Flag Field F1-F7 bits establish connection, including the security code for recognition in mobile-to-base phone interconnects. Such bits are in don't care states during other calls.

When the originating transmitter has completed the foregoing, the operator presses a "send" button to do the following: the radio will look for an idle local talk channel. If there is none, it will look for an idle repeater talk channel; and busy out if there isn't one. If there is a local talk channel free, the talk channel field will be set in the control message, the Flag Field bit to F0, and a control message is transmitted to the ultimate receiver. The caller then listens to the low band of the control channel for acknowledgment of the message. If this is received within 2.2 seconds the talk channel field is setup. If not, there is a random delay of 2 seconds, and then the initial scan attempt is tried again. Four attempts, all ending in failure, will cause the caller to busy out.

As an acknowledgment is received, the transmitter examines the return message and channel number for a match and then lockup. A different channel suggestion will cause the transmitter to examine the change, and if it is busy, will begin a new search after a random delay of two seconds. If OK, the message originator will set the Talk Channel Field, etc. with the new channel and await a response.

Should a repeater talk channel be chosen, the message originator will set the Talk Channel Field and the Flag Bit to F1. A control message then goes to the receiver through this repeater on the low band control channel. Meanwhile, the repeater error checks, reclocks, and resynchronizes all control contents, then assigns a repeater talk channel.

The transmitter listens to the high band control frequencies of the control channel for message receipt, and if all is well, the transmitter and receiver go in step. Otherwise, action will revert to the repeater for additional channel assignment. When the transmitter is satisfied with the receiver's reply, it will proceed to the channel assigned and go into lock step. For calls other than mobile-to-phone lines, a ringing sidetone can be expected at the receiver's station. For mobile-to-telephone line calls, a dial tone is heard on the talking channel. Following a 1-second delay, dialed digits will be sent for the phone call.

PRCS Receiving Rules differ, of course, from those of the

message originator or transmitter. All steps apply to mobile and base stations alike.

As a message in the Destination Address field is received, the receiver decodes control message fields and checks for the following:

☐ If flag F1 from mobile or base and F2-F7 don't match a code input into the base, then the control message is ignored and the receiver continues monitoring.

☐ But if the message transmitted is accepted, then it is stored for later use. And if F0 is a 1, then a repeater call is incoming, and the repeater designates a talk channel. But if Flag F0 is 0, a local channel appears and must be channel-checked. If this channel is available the control message will be acknowledged and include the Source Address Field, the Destination Field, and other fields must be equivalent. The receiver then goes to the talk channel pair suggested. If this particular channel is busy, the receiver will scan local talk channels and either find one or return to monitoring.

☐ If an idle channel is available, the control message will be acknowledged and a number for the idle channel transmitted. Afterwards, the receiver will return to monitoring, awaiting either an additional message or handshake to begin the conversation.

If a station-to-station call is suggested by F1, the receiver will generate a ringing sidetone heard by the caller, then automatically interconnect the base station to telephone lines and send a dial tone signal to the transmitting station which produces the proper phone number.

Disconnect and/or Default. Each station will be able to automatically disconnect ongoing calls after 100 seconds, ensuring an equal opportunity to share the service by all. And during peak usage hours, station disconnects can be overridden by repeater disconnects after 100 seconds during morning and evening rush hours. There is also a default mechanism available so that both stations can terminate any call and release their talk channel under these following conditions:

☐ Where a call is begun on a local channel, ringing ensues, but the initiator moves out of range before any conversation. Ringing could continue occupying this particular channel, but in the absence of rf carriers from each transceiver, the two units cease all contact within 10 seconds.

☐ A brief conversation occurs and ends, but the originator never hangs up, although the message recipient does. Default occurs again within 10 seconds since one of the parties has stopped its transmissions.

☐ Two stations again converse, but the conversation initiator refuses to hang up even though the other party does. Once more, the 10-second default mechanism shuts down both transceivers.

Upon receipt of a disconnect signal, the normal repeater high band talk channel repeats it, the originating station disconnects and the receiving station then disconnects upon command freeing the channel for further use.

When system loads exceed 70 percent, a 1-second tone burst at 1150 Hz with full modulation will warn subscribers that a timeout follows in 9 seconds. Otherwise, the timer resets and there is no warning as long as channel occupancy does not exceed 70 percent.

Idle Receive. The five control channels are to be monitored constantly for control messages, looking for word sync in the command message preamble. Command messages addressed to a certain repeater are stored and the repeater enters the Command Message Process task. Should the message have another address, the Source Flag word has to be checked for bit 1. If set, the Destination Response Search task is entered; if not set, the task is to be reentered. Should the Reply Timer timeout during the operation, the repeater signifies Call Failed. Initially, when power is applied, the repeater enters initialization and a valid subscriber list is stored in memory, timers are set to zero, and channel availability checked. It immediately looks for a 2850 Hz supervisory tone signifying channel use. If occupied, the repeater starts a 1.5 minute timer, monitoring the channel for disconnect. With no occupancy noted within 1 second, the repeater sends a disconnect signal to source and destination units, disconnects, and reenters Idle Receive.

Repeaters

Generally, the function of any PRCS repeater is to accept signals from one station and transfer them to another station within some distance of 10 to 20 miles, depending on power out, antenna height, gain, and local terrain. It must accurately determine if the incoming signal originates from a valid subscriber, find an available open channel, and connect the two talking stations. The repeater does the following:

☐ Receives and stores in memory the calling control message until released on a clear channel.

☐ Verifies the caller as a legitimate subscriber.

- ☐ Monitors available channels.
- ☐ Matches an idle channel to the call while observing calling channel choice, if possible.
- ☐ Retransmits control signals to the distant receiver.
- ☐ Returns acknowledgment from receiver to caller, with message reception on low band and retransmission on high band.
- ☐ When 70 percent of the repeater channels are occupied, it terminates the various communications through the repeater within 60 seconds after inception.

G.E. says that computer technology is to be the prime electronics of repeaters, furnishing control, logic, and all memory. Such memory is programmed with an up-to-date subscriber listing and logic checks this against any incoming control message. Logic also maintains a used and unused channel listing for immediate message transfer to waiting receivers. Input channels will also be checked to make sure they are not being used by some adjacent repeater. Further, transmitter signals are not relayed unless a subscriber is listed for that particular repeater. This is done to reduce interference, maintain proper storekeeping, and prevent two repeaters from accepting and transmitting the same control or other message. This is why handshaking between transmitter and repeater is so important.

Power Cutbacks and Control

To prevent carrier and signal interference between communicating stations, base and mobiles are equipped with sensing devices to reduce rf power. Unlike cellular systems where gain reductions are centralized in a specific controller, PRCS stations can sense all received signals and decrease power accordingly.

If an initially received signal is greater than -60 dBm ± 5 dBm, transmitter power output will be reduced to a level of -5 dBW or less. Under -60 dBm, transceiver output power level remains at $+10$ dBW max. Previously, G.E. had planned to reduce power in one or two 15 dB steps, depending on attenuation needed when incoming signals were 35 dB above a receiver's 12 dB SINAD sensitivity level. This would have corresponded to a 25 dB reduction in received power over initial inputs based on an averaging time over the period of sensing. Control signals, says G.E., are not affected since such transmissions are short and maximum signal strength in digital signaling to prevent errors is always desirable.

Repeaters are not affected either in any rf power reduction action since they are not reused locally and there is nothing to gain.

Consequently, repeaters operate at all times under full power for maximum efficiency and strong message transmissions.

PROPOSED PARAMETERS

Parameters suggested by General Electric for the Personal Radio Communications Service are listed in sequential order for available channels and logically for the remainder. Many, of course, have already been covered and most of these will not be repeated since text for the chapter is relatively short and paragraph headings explicit. Again, these statistics are based on 149/150 channel pairs rather than 133 as more recently proposed since the FCC has made no final decision on G.E.'s request during at least the first half of 1983. Therefore, consider them for what they are; proposals for a brand new 900 MHz service that won't emerge until at least 1984 or later.

Usage of the channels in Table 5-1 breaks down as follows: Repeater station licensees may operate on channels 1-5 and 39-149. Channels 1-5 are designated Control Channels and through them is determined the channel pairs to be used during normal voice communications. Channels 6-37 are reserved for selective calling between two mobiles or a base and another mobile station, and are designated Local Talk Channels.

Channels 38-149 are the Repeater Talk channels for use in selective calling communications between two mobiles or between a mobile and base station via Repeaters. Channel 150 (low and high) is/are for nonaddressable, direct communications between mobiles or base stations and are called Party Line Channels for everyone's use. Channel bandwidth, as you will recall, is 30 kHz total without guard bands.

Table 5-1. Proposed Personal Radio Communications Channels in Pairs.

Channel No.	Frequency Pairs in MHz
1	887/932
2-5	887.030/932.030
6	887.150/932.150
7	887.180/932.180
. . . .	
37	888.080/933.080
38	888.110/933.110
39	888.140/933.140
. . . .	
148	891.410/936.410
149	891.440/936.440
150	891.470/936.470

Repeater Station Specifications

These are not to exceed the heights and the effective radiated power (ERP) specified:

0-400 feet	100 W
400-800 feet	60 W
800+ feet	30 W

Emissions

Types F3 and F9 are suggested for voice and data transmissions, with peak frequency deviations to ±8 kHz and ±5 kHz, respectively. Signals between 3 kHz to 20 kHz by at least 60 log (f/3) dB, relative to 1 kHz attenuation and f= 1 kHz. Signals above 20 kHz are reduced by at least 50 dB, relative to the same attenuation.

Antennas

PRCS stations are to have vertically polarized antennas.

Audio Tone

PRCS base and mobile stations operating on local talk or repeater channels will modulate a continuous 2.850 kHz ±5 Hz tone with 0.8 kHz deviation.

Timer-Disconnects

Base and mobile stations are to be equipped with timers that disconnect and cut off communications on channels 6-149 automatically after 180 seconds and also disconnect any base from the telephone network. And repeater stations shall disconnect any or all communications when their duration reaches 100 seconds during periods when 70 percent of all Repeater Talk Channels are in use. Also, when local or repeater talk channels receive no signals during normal conversations for 10 seconds, they are to terminate use of the particular channel.

Control Channels

During standby, base and mobile stations monitor a control channel identified numerically with their station address:

Channel	Address
1	000-199

Channel	Address
2	200-399
3	400-599
4	600-799
5	800-999

Control messages begin with the Preamble, Start, source Address, Destination, Talk Channel, Flag, and end with CRC-16. The digital signal rate transmits at 10 kilobits/second; data is modulated over control or talk channels by minimum shift keying of the carrier.

Fortunately, Signaling Formats, as submitted to the FCC by G.E. in a late response are available and are included, hereafter, just as received (Courtesy General Electric Co., PRCS Systems Engineering, Syracuse, N.Y.) They should offer substantial aid in further understanding the system.

SIGNALING FORMATS

These are included as received immediately after transmission to the Federal Communications Commission and included with slight editing.

Command Message Format

All command messages that are sent over a control channel consist of a 560 bit packet arranged in the manner shown. The left-most bit is earliest in time, and shall be designated the most significant bit.

MSB   LSB

DOTTING	WORD SYNC	MESSAGE FRAME SENT SEQUENTIALLY THREE TIMES
97	31	432

PRECURSOR

```
DOTTING    = 010101 . . . 0
WORD SYNC  = 1111100011011101010000100101100
```

The command messages begin with a 128 bit precursor which consists of a 97 bit dotting sequence followed by a 31 bit word sync sequence. The precursor is followed by the message frame which is sequentially transmitted three times within the command message packet. The total control channel message length is therefore 128 bits (precursor) plus 3 × 144 bits (message frame), or 560 bits.

Message Frame Format

Each 144 bit message frame consists of four fields arranged in the manner shown.

MSB   LSB

DESTINATION ADDRESS FIELD	SOURCE ADDRESS FIELD	CONTROL FIELD	CRC-CCITT
40	40	48	16

The first 40 bits of the message frame defines the destination address field. This is followed by a 40 bit source address field, a 48 bit control field, and a 16 bit CRC-CITT field. The details of the respective fields will now be described.

Destination Address Field. The destination address field is defined as follows:

MSB   LSB

DEST FLAG	DESTINATION IDENTIFICATION NUMBER	DESTINATION UNIT NUMBER
2	34	4

☐ The destination address flag word is a two bit word that identifies whether the command message is from the calling unit, or from the called unit that is responding to a command message involving a direct talk channel. The individual bits are defined as follows:

DF1 - set to 1 when the command message is from a called unit that is responding to a direct talk channel (DTC) command message, cleared to 0 otherwise.
DF0 - not defined.

☐ The destination identification number word has 34 bits which defines the identification number of the final destination of the call or an action command to the destination unit. The destination may be either another radio unit, or a telephone that is called via the public switched telephone network through a home station. Provision is therefore made to include an area code plus a seven digit telephone number in this word. The format is defined as follows:

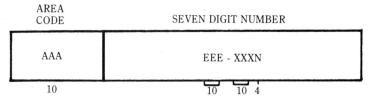

AAA 3 decimal digits encoded as 10 bit binary word
EEE 3 decimal digits encoded as 10 bit binary word
XXX 3 decimal digits encoded as 10 bit binary word
N   1 decimal digit encoded in BCD format

The format and meaning of the Destination Identification Number word, when used to represent an action command to the destination unit, is not defined.

☐ The destination unit number word is a four bit, consumer supplied, identifier that represents one hexadecimal digit and allows sixteen units to be addressed with the same identification number. For a telephone call from a mobile transceiver, the Destination Unit Number is the unit number of the home station which interfaces the public switched telephone network (PSTN). The address of said home station is constructed from the source identification number and the destination unit number.

Source Address Field. The source address field is defined as follows:

SOURCE FLAG	SOURCE IDENTIFICATION NUMBER	SOURCE UNIT NUMBER
2	34	4

MSB ... LSB

☐ The source flag word is a two bit word that identifies whether the command message is from a calling unit or from the called unit responding to a command message involving a repeater talk channel. The bits are defined as follows:

SF1 - Set to 1 when the command message is from a called unit that is responding to a repeater talk channel (RTC) command message, cleared to 0 otherwise.
SFO - Not Defined.

☐ The source identification number word has 34 bits in the same format as the destination identification word.

☐ The source unit number word is a four bit, consumer supplied, identifier that represents one hexadecimal digit and allows sixteen units to be addressed with the same identification number.

Control Field. The control field is a 48-bit field that is structured in the manner shown.

MSB			LSB
TALK CHANNEL	VALIDATION CODE	DAD	STATUS
8	32	4	4

☐ The talk channel word is an 8-bit word which identifies the talk channel that the sending unit wishes to use. Table 5-2 details the meaning of the respective codes. Codes 0 0 0 0 0 0 0 0 through 0 0 0 1 1 1 1 1 identify one of thirty-two direct talk channels. Codes 0 0 1 0 0 0 0 0 through 0 1 1 1 1 1 1 1 are repeater identification codes and are used by the calling unit to address one of 96 possible repeaters. Codes 1 0 0 0 0 0 0 0 through 1 1 1 0 1 1 1 1 identify one of 112 repeater talk channels. Codes 1 1 1 1 0 0 0 1 through 1 1 1 1 1 1 1 1 are presently unassigned. Code 11110000 is used as a null channel designator when the called unit is responding with a status message that will disrupt the command protocol.

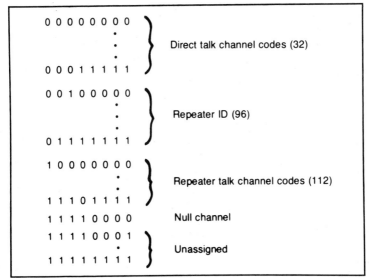

Table 5-2. Talk Channel Word Coding. (Courtesy General Electric)

Table 5-3. Destination Address Data (DAD) Word Coding. (Courtesy General Electric)

MSB LSB	
0 0 0 0	Direct addressing (radio-to-radio call)
0 0 0 1	Seven digit telephone number
0 0 1 0	Ten digit telephone number (AC + 7 digits)
0 0 1 1	Eleven digit telephone number (1 + AC + 7 digits)
0 1 0 0	Eleven digit telephone number (0 + AC + 7 digits)
0 1 0 1 ↓ 1 1 1 0	Unassigned
1 1 1 1	Action command message

☐ The validation code word is a 32-bit word which is used to restrict telephone network access through a home station by a mobile unit. It consists of two parts; a 28-bit Transceiver Identification Number (TIN), which is a 7 digit hexadecimal number assigned by the manufacturer and permanently stored in the transceiver's memory, and a 4-bit privacy code that is set by the consumer. The structure of this word is as follows.

MSB	LSB
PRIVACY CODE	TRANSCEIVER IDENTIFICATION NUMBER (TIN)
4	28

☐ The Destination Address Data (DAD) word is a 4-bit word that is used to define the meaning of the Destination Identification Number. The coding is defined in Table 5-3. All zeros are sent when the calling unit is not making a telephone call, and the identification number used in the destination address field is the identification number of the unit being called. The remaining assigned codes, exclusive of 1111, pertain to telephones that are reached by a call on the public switched telephone network and identify the number telephone digits that have been sent. Shorter length numbers may be accommodated by ignoring leading zeros within the specified length. Code 1111 is sent when the information sent in the Destination Identification Number word is an action command word of 34 bits that the destination should execute.

☐ The status word is a four bit word in which the most significant bit is unassigned. Coding of the three remaining bits is defined in Table 5-4.

CRC-CCITT Field. This field is a 16 bit word that is used for error detection in the Message Frame. The generator polynomial is $X^{16} + X^{12} + X^5 + 1$.

Command Messages Index

The use of the signalling format produces many different messages to be sent over a control channel. This section delineates such messages as used in the PRCS protocol. Following the index are the message frame breakdowns by command message.

Unit to Unit Over Direct Talk Channel (DTC)

1. UT   - Direct Talk Channel Request
2. UTU - Called Unit Response to Direct Talk Channel Request w/Agreement on DTC.
3. UTU - Called Unit Response to Direct Talk Channel Request w/new DTC
4. UTU - Called Unit Response to Direct Talk Channel Request w/no DTC Available

Mobile Telephone Call Over Direct Talk Channel

5. UTUT - Direct Talk Channel Request
6. UTUT - Home Station Response to Direct Talk Channel Request w/Agreement on DTC.
7. UTUT - Home Station Response to Direct Talk Channel Request w/new DTC
8. UTUT - Home Station Response to Direct Talk Channel Request w/no DTC Available
9. UTUT - Home Station Response to Direct Talk Channel Request w/Telephone Busy
10. UTUT - Home Station Response to Direct Talk Channel Request w/Validation Code Error

Table 5-4. Status Word Code Definition.

MSB LSB	
X 0 0 0	Everything OK
X 0 0 1	No channel available
X 0 1 0	
X 0 1 1	Called unit did not reply to repeater
X 1 0 0	New channel
X 1 0 1	Validation code error
X 1 1 0	Repeater nonsubscriber disconnect
X 1 1 1	Telephone in use (at interface)

Unit to Unit Through Repeater Talk Channel (RTC)

11. URU - Repeater Talk Channel Request
12. URU - Repeater to Called Unit
13. URU - Called Unit Response w/Agreement on Repeater Talk Channel
14. URU - Repeater Response to Calling Unit w/no Repeater Talk Channel Available
15. URU - Repeater Response to Calling Unit w/no Response from Called Unit
16. URU - Repeater Response to Calling Unit who is not a Subscriber

Mobile Telephone Call Through a Repeater

17. URUT - Repeater Talk Channel Request
18. URUT - Repeater to Called Unit
19. URUT - Home Station Response w/RTC OK
20. URUT - Repeater Response to Calling Unit w/no RTC Available
21. URUT - Repeater Response to Calling Unit w/no Response from Home Station
22. URUT - Repeater Response to Calling Unit who is not a Subscriber
23. URUT - Home Station Response w/Telephone Busy
24. URUT - Home Station Response to RTC Request w/Validation Code Error

☐ Unit-to-Unit (UTU): Direct Talk Channel (DTC) Request

Destination Address:
FLAG: OX
Area Code: Called Unit Area Code
Identification Number: Called Unit Identification Number
Unit Number: Called Unit ID

Source Address:
FLAG: OX
Area Code: Calling Unit Area Code
Identification Number: Calling Unit Identification Number
Unit Number: Calling Unit ID

Control Field:
Talk Channel: Requested Channel Code
 (00000000 to 00011111)
Validation Code: $P_3 P_2 P_1 P_0\ T_{27} - T_0$ (Calling)
DAD: 0000
Status: X000

☐ Unit-to-Unit (UTU): Response Direct Talk Channel w/ Agreement

Destination Address:	FLAG: 1X Area Code: Calling Unit Area Code Identification Number: Calling Unit Identification Number Unit Number: Calling Unit ID
Source Address:	FLAG: 0X Area Code: Called Unit Area Code Identification Number: Called Unit Identification Number Unit Number: Called Unit ID
Control Field:	Talk Channel: Channel Code Requested by Source Validation Code: $P_3P_2P_1P_0$ $T_{27}-T_0$ (Called) DAD: 0000 Status: X000

☐ Unit-to-Unit (UTU): Response to Direct Talk Channel Request w/Disagreement - New Channel in Response

Destination Address:	FLAG: 1X Area Code: Calling Unit Area Code Identification Number: Calling Unit Identification Number Unit Number: Calling Unit ID
Source Address:	FLAG: 0X Area Code: Called Unit Area Code Identification Number: Called Unit Identification Number Unit Number: Called Unit ID
Control Field:	Talk Channel: New Direct Talk Channel Code Validation Code: $P_3P_2P_1P_0$ $T_{27}-T_0$ (Called) DAD: 0000 Status: X100

☐ Unit-to-Unit (UTU): Response to Direct Talk Channel Request w/Disagreement - No Direct Talk Channel Available at Called Unit

Destination Address:	FLAG: 1X Area Code: Calling Unit Area Code Identification Number: Calling Unit Identification Number

Source Address:	Unit Number: Calling Unit ID FLAG: 0X Area Code: Called Unit Area Code Identification Number: Called Unit Identification Number
	Unit Number: Called Unit ID
Control Field:	Talk Channel: 11110000 Validation Code: $P_3 P_2 P_1 P_0\ T_{27} - T_0$ (Called) DAD: 0000 Status: X001

☐ Unit-to-Unit-to-Telephone (UTUT): Mobile Telephone Call - Direct Talk Channel Request

Destination Address:	FLAG: 0X Area Code: Telephone Area Code Identification Number: Telephone Number Unit Number: Home Station Unit ID
Source Address:	FLAG: 0X Area Code: Calling Unit Area Code Identification Number: Calling Unit Identification Number Unit Number: Calling Unit ID
Control Field:	Talk Channel: Direct Talk Channel Requested Validation Code: $P_3 P_2 P_1 P_0\ T_{27} - T_0$ (Calling) DAD: Function of Destination Telephone Number (Not 0000) Status: X000

☐ Unit-to-Unit-to-Telephone (UTUT): Direct Talk Channel Mobile Telephone Call - Response w/Telephone Busy

Destination Address:	FLAG: 1X Area Code: Calling Unit Area Code Identification Number: Calling Unit Identification Number Unit Number: Calling Unit ID
Source Address:	FLAG: 0X Area Code: Called Unit Area Code Identification Number: Called Unit Identification Number Unit Number: Called Unit ID

Control	Talk Channel: 11110000
Field:	Validation Code: P_3-P_0 $T_{27}-T_0$ (Called)
	DAD: 0000
	Status: X111

☐ Unit-to-Unit-to-Telephone (UTUT): Direct Talk Channel Mobile Telephone Call - Response with Validation Code Error

Destination	FLAG: 1X
Address:	Area Code: Calling Unit Area Code
	Identification Number: Calling Unit Identification Number
	Unit Number: Calling Unit ID
Source	FLAG: 0X
Address:	Area Code: Called Unit Area Code
	Identification Number: Called Unit Identification Number
	Unit Number: Called Unit ID
Control	Talk Channel: 11110000
Field:	Validation Code: P_3-P_0 $T_{27}-T_0$ (Called)
	DAD: 0000
	Status: X101

☐ Unit-to-Repeater-to-Unit (URU): Repeater Talk Channel (RTC) Request (To Repeater)

Destination	FLAG: 0X
Address:	Area Code: Called Unit Area Code
	Identification Number: Called Unit Identification Number
	Unit Number: Called Unit ID
Source	FLAG: 0X
Address:	Area Code: Calling Unit Area Code
	Identification Number: Calling Unit Identification Number
	Unit Number: Calling Unit ID
Control	Talk Channel: Repeater ID (00100000)→(011111111)
Field:	Validation Code: P_3-P_0 $T_{27}-T_0$ (Calling)
	DAD: 0000
	Status: X000

☐ Unit-to-Repeater-to-Unit (URU): Repeater to Called Unit

Destination Address:	FLAG: 0X Area Code: Called Unit Area Code Identification Number: Called Unit Identification Number Unit Number: Called Unit ID
Source Address:	FLAG: 0X Area Code: Calling Unit Area Code Identification Number: Called Unit Identification Number Unit Number: Calling Unit ID
Control Field:	Talk Channel: Repeater Talk Channel (10000000) → (11101111) Validation Code: $P_3-P_0\ T_{27}-T_0$ (Calling) DAD: 0000 Status: X000

☐ Unit-to-Repeater-to-Unit (URU): Called Unit Response w/ Agreement on Repeater Talk Channel.

Destination Address:	FLAG: 0X Area Code: Calling Unit Area Code Identification Number: Calling Unit Identification Number Unit Number: Calling Unit ID
Source Address:	FLAG: 1X Area Code: Called Unit Area Code Identification Number: Called Unit Identification Number Unit Number: Called Unit ID
Control Field:	Talk Channel: Repeater Talk Channel Validation Code: $P_3-P_0\ T_{27}-T_0$ (Called) DAD: 0000 Status: X000

☐ Unit-to-Repeater-to-Unit (URU): Repeater Response to Calling Unit w/No Repeater Talk Channel Available (at Repeater).

Destination Address:	FLAG: 0X Area Code: Calling Unit Area Code Identification Number: Calling Unit Identification Number Unit Number: Calling Unit ID

Source Address:	FLAG: 1X Area Code: Calling Unit Area Code Identification Number: Calling Unit Identification Number Unit Number: Calling Unit ID
Control Field:	Talk Channel: Repeater ID Validation Code: 0101010101010101 DAD: 0000 Status: X001

☐ Unit-to-Repeater-to-Unit (URU): Repeater Response to Calling Unit w/No Response from Called Unit

Destination Address:	FLAG: 0X Area Code: Calling Unit Area Code Identification Number: Calling Unit Identification Number Unit Number: Calling Unit ID
Source Address:	FLAG: 1X Area Code: Called Unit Area Code Identification Number: Called Unit Identification Number Unit Number: Called Unit ID
Control Field:	Talk Channel: Repeater ID Validation Code: 0101010101010101 DAD: 0000 Status: X011

☐ Unit-to-Repeater-to-Unit (URU): Repeater Response to Non-Subscriber Calling Unit

Destination Address:	FLAG: 0X Area Code: Calling Unit Area Code Identification Number: Calling Unit Identification Number Unit Number: Calling Unit ID
Source Address:	FLAG: 1X Area Code: Called Unit Area Code Identification Number: Called Unit Identification Number Unit Number: Called Unit ID
Control	Talk Channel: Repeater ID

Field: Validation Code: 0101010101010101
DAD: 0000
Status: X110

☐ Unit-to-Repeater-to-Unit Mobile Telephone Call (URUT): Repeater Talk Channel Request

Destination FLAG: 0X
Address: Area Code: Telephone Area Code
             Identification Number: Telephone Number
             Unit Number: Home Station Unit ID
Source   FLAG: 0X
Address: Area Code: Calling Unit Area Code
             Identification Number: Calling Unit
                                    Identification Number
             Unit Number: Calling Unit ID
Control   Talk Channel: Repeater ID
Field: Validation Code: $P_3-P_0\ T_{27}-T_0$
             DAD: Function of Telephone Number
             Status: X000

☐ Unit-to-Repeater-to Unit Mobile Telephone Call (URUT): Repeater to Home Station

Destination FLAG: 0X
Address: Area Code: Telephone Area Code
             Identification Number: Telephone Number
             Unit Number: Home Station Unit ID
Source   FLAG: 0X
Address: Area Code: Calling Unit Area Code
             Identification Number: Calling Unit Identification
                                    Number
             Unit Number: Calling Unit ID
Control   Talk Channel: Repeater Talk Channel (RTC)
Field: (10000000) → (11101111)
             Validation Code: $P_3-P_0\ T_{27}-T_0$
             DAD: Function of Telephone Number
             Status: X000

☐ Unit-to-Repeater-to-Unit Mobile Telephone Call (URUT): Called Unit Response w/Telephone Busy

Destination Address:	FLAG: 0X Area Code: Calling Unit Area Code Identification Number: Calling Unit Identification Number Unit Number: Calling Unit ID
Source Address:	FLAG: 1X Area Code: Called Unit Area Code Identification Number: Called Unit Identification Number Unit Number: Called Unit ID
Control Field:	Talk Channel: 11110000 Validation Code: $P_3, P_2, P_1, P_0\ T_{27}-T_0$ (Called) DAD: 0000 Status: X111

☐ Unit-to-Repeater-to-Unit Mobile Telephone Call (URUT): Home Station Response with Validation Code Error

Destination Address:	FLAG: 0X Area Code: Calling Unit Area Code Identification Number: Calling Unit Identification Number Unit Number: Calling Unit ID
Source Address:	FLAG: 1X Area Code: Called Unit Area Code Identification Number: Called Unit Identification Number Unit Number: Called Unit ID
Control Field:	Talk Channel: 11110000 Validation Code: $P_3, P_2, P_1, P_0\ T_{27}-T_0$ (Called) DAD: 0000 Status: X101

Chapter 6

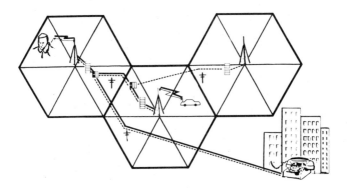

Test Programs and Equipment

Yes, the title of this chapter means precisely what it says, but we don't yet know exactly what is meant. If that sounds like a juvenile riddle, it isn't! Cellular communications test programs and test sets are still in development and, undoubtedly, will continue in this vein for some time to come. Of course there will be individual 2-way radio testers, some with plug-ins that will adapt to general mobile as well as specific cellular. . . Cushman Electronics is an outstanding example since their Model 4000 monitor could be designed to have this capability.

On the other hand, General Electric and Northern Telecom are working on remote diagnostics from the Switch that can check out at least every major parameter of both cell sites and the mobiles themselves, including frequency tolerance and receiver sensitivity. So while good test sets with nanovolt sensitivity will continue to be quite handy for pagers and run-of-the-mill mobiles, signaling operations as well as radio response are to become dual factors in all cellular servicing precedures that will require looking at 666 channel tolerances as tight as 1.5 ppm in cell sites and 5 ppm for mobiles.

Guaranteed monthly or annual oscillator drift in test set rf stages, then, is to become highly important unless you want to calibrate your equipment almost every time you use it—a rather expensive practice if your gear is subject to unusual frequency intolerance. Should this be the problem, even a fairly expensive TCXO addition and oscillator rework might save a lot of time and some money in the end. If this can't be done, then a new test set should be immediately considered. And in order to measure such parameters accurately, the measuring units should have at least double the accuracy of the radio or installation being measured.

TEST EQUIPMENT REVIEW

Just to put this whole measuring process in perspective and point out some differences to be expected between 25-40 MHz and 400-900 MHz, let's review some oscilloscope and spectrum analyzer applications among the lower frequencies first before tackling the coming esoterics. I fully realize that black box readouts of sensitivities, SWRs, frequency tolerances, power output, plus spurs and harmonics are the coming thing and could well appear over the next several years. I also know that to look at any rf responses among higher spectrum carriers will require GHz oscilloscopes and spectrum analyzers. But we've all got to start somewhere, and a little low frequency eyewetting isn't going to hurt a bit, other than prepare you for some occasion when you're not servicing cellular or have lost your cherished autoreadout gadgets and have only conventional equipment handy. So with those words of incalculable wisdom freshly laid on paper, let's get down to the serious business of what must be used.

Oscilloscopes

Usually for ordinary servicing purposes, a 20 MHz, dual trace, single time base unit with 3-5 percent overall accuracy works out just fine. Fortunately, with so many of the manufacturers featuring 100 MHz units at $2,000 or more, state-of-the-art and fierce price competition has actually reduced the retail price for more modest equipments to under $1,000, offering very usable instruments with included probes in the package. Unfortunately, however, most of these low-priced, high-utility oscilloscopes are pure and simple imports, and you have to watch quality and certain specifications like a hungry eagle. Some are totally unserviceable and others barely make their designated parameters, allowing no room at all for component aging, catastropic conditions, or even ordinary storage in hot or cold environments. This, of course, means numerous calibrations and considerably more service than normal.

When buying a scope, it should be checked by someone with reasonable familiarity to be sure it will pass muster. You'd be surprised what can be discovered in a half hour of frequency handling, sync, and calibration checks. Such an examination won't guarantee special instrument longevity or freedom from down time, but there is no substitute for a careful head start. Here are some examples:

☐ Variac ac power input between at least 110 Vac and 130 Vac. The better scopes will hold parameters to at least 107 volts on the low end, and then they permit a transformer tap adjustment for ranges different from 120 Vac or 240 Vac.

☐ Use a good sinewave or, better yet, a triangular wave generator to examine time base calibration between 100 Hz and 1 MHz, connecting a counter to an accompanying splitter, if you wish. A 5 percent scope should calibrate to within 4 percent or less, and a 3 percent scope to 2 percent overall.

☐ With dc voltage or a constant voltage generator offering square waves, examine the two vertical amplifiers the same way. Both should calibrate accurately throughout their range, which should extend from at least 5 mV to 20 V/division. Good time bases stretch from at least 0.5 sec to 500 nsec, with a times 5 or 10 expander.

☐ You also want to be sure there is sufficient accelerating potential at the cathode ray tube to display with good visibility high frequencies to 20 MHz and beyond. And in less than brightly lighted work areas, an oscilloscope with adequately lighted graticule is a must. Otherwise you'll suffer continual eyestrain from trying to read trace positions, amplitudes, and waveform content. For a 20 MHz scope you'll need not less than 4 kV and preferably 6 kV or better between cathode and anode to supply sufficient trace illumination. At 25 MHz and above, at least 10 kV becomes minimum. Whatever the 3 dB down bandwidth, excellent voltage regulation is required for satisfactory oscilloscope operation.

☐ Triggering. Either channel should trigger on simple or complex waveforms across the frequency range aided, perhaps, by manual or automatic selection of low or high pass filters. Further, on tv vertical and horizontal repetiton rates such as 59.94 Hz and 15,734 Hz, full sync should be available in the automatic mode without manual filter introduction, if possible. Sometimes, however, a 60 Hz filter means the difference between positive sweep lock and a shaky trace—the latter being impossible to tolerate. Normally, sine waves and square waves are no problem for even the lesser expensive models. But there are, even now, new scopes on the market that will not auto sync 15-16 kHz. Try on a tv receiver using standard off-the-air signals before you buy if there is any question at all.

Time Bases

For some reason, this highly important portion of any oscilloscope presents something of an enigma to traditional service personnel in almost any trade. Here, as Franklin Roosevelt was fond of saying, "there's nothing to fear but fear itself." For you have little more than a sweep generator offering either AT (automatic triggering) or NT (normal triggering), with the incoming waveform supplying the synchronizing edge. Time constants marked in seconds, milliseconds, and microseconds are usually the time intervals seen, although nanoseconds (10^{-9}) do appear on the more expensive models. However, as frequency handling abilities increase, the vertical deflection factor decreases, and high frequency oscilloscopes often have no more vertical range than 2 millivolts to 5 volts per centimeter or division. Usually this is quite insufficient for routine servicing, being reserved primarily for solid state and other allied low voltage applications. Ultra high frequencies and large vertical deflections are normally not partners, at least among oscilloscope manufacturers. Probably the best example of such an instrument that I have seen was Telequipment's 25 MHz D67A, which had an overall vertical deflection of from 1 mV to 50 V/division. The D67A block diagram is shown in Fig. 6-1. Unfortunately, the English manufacturer may no longer exist, but parts continue to be available through Tektronix depots around the country.

Any worthwhile scope should trigger on signals half a division in graticule height, some less. On low frequency signals, there is a chopped mode where a free-running oscillator between 250 kHz and 1 MHz will lock both traces on the screen, selecting samples of each for simultaneous display. If switching transients don't interfere, this mode is good up to the low milliseconds for suitable dual-trace waveform comparisons. After that, you had best use the alternate switching mode for first one vertically amplified trace and then the other. As sweep speeds increase, the two traces appear more and more in time, and so you are led to believe they appear simultaneously, just like chopped. But that isn't so since the sweep is waveform-triggered and first one channel is presented and then another.

To aid in such a presentation, there are a few helpers available that should be parts of all standard oscilloscopes. Once the meaning of these aids is understood, there should be little problem in using

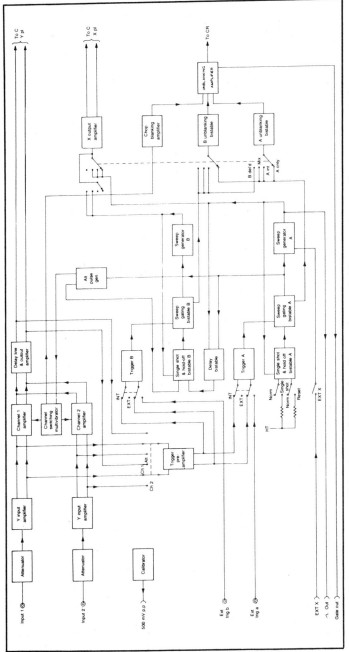

Fig. 6-1. Block diagram of older Telequipment D67A. It is still a good example of an excellent oscilloscope currently in use.

them if there is enough applied practice. Let's define a few of these for good measure.

Trigger Level, in combination with positive or negative slope, permits sweep triggering from any portion of the signal edge, including those of complex waveshapes. This, of course, is in the Normal mode. In the Automatic mode, the trigger level assumes a much broader range—virtually from on to off—and the sweep generator is triggered automatically, even supplying a baseline trace when there is no incoming signal. Here the time base is set to the approximate time period involved and will lock on any well-defined signal whose rise and fall times are found broadly suitable.

Trigger Indicators on some oscilloscopes in the form of light emitting diodes (LEDs) will illuminate when the time base is triggered and ready to deliver a stable display. On slow signals this can aid in adjusting the Level control during Normal operation.

Sweep Delays usually permit some portion of the waveform to be selected from another at some specific time so that it may be expanded and examined in detail. There are several ways to do this, ranging from two separate time bases to a psuedo-delay that you may find interesting and possibly entirely suitable to your needs. We'll briefly describe both.

In the double time base arrangement, a portion of time base A reaches time base B, turning it on for some period determined by time base B control setting. Always slower than time base A, the waveform is displayed as an expanded portion of A, with considerably more detail evident. Initially, the information to be displayed appears as an intensified part of the trace with no expansion. Thereafter, the delayed portion may either be shown in a mixed sweep (delayed and non-delayed) mode, or in full delay without an initial portion of the waveform in evidence. I have found mixed sweep very handy (Fig. 6-2) especially where there were complex waveforms that could be peeled off by a calibrated 10-turn potentiometer selecting precise portions of the trace.

Psuedo-sweep, illustrated in Fig. 6-3, amounts to a simple delay in the start of any sweep between 100 nanoseconds and 1 second so that some portion of the display may be selected for expansion by either changing the sweep speed or using the X10 magnifier, or both. Of course the trace dims somewhat as speeds become super fast, and you'll have to choose some optimum viewing set for either general or specific purposes. Naturally, the dual time base with its sampling and trigger circuits is much the more expen-

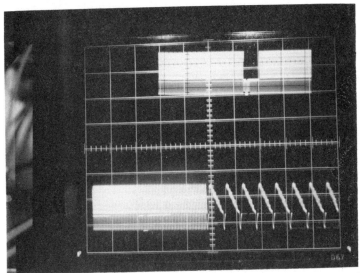

Fig. 6-2. Dual photos of psuedo- and real-delayed sweeps, including mixed sweep.

sive, but mixed sweep is very appealing, nonetheless. Your choice should depend exclusively on the application, always kept within limitations of the particular oscilloscope you're selecting. The HM 204 Hameg shown in Fig. 6-4 also has a normal, search, and delay

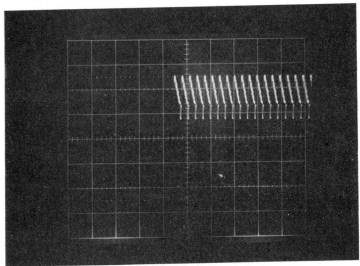

Fig. 6-3. Delayed portion of Hameg pseudo-sweep.

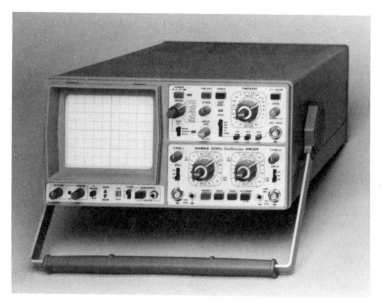

Fig. 6-4. The Hameg HM 204 dual trace, 20 MHz servicing oscilloscope. It's fully recommended.

switch, accompanied by a LED indicator denoting functional engagement. A 20-turn potentiometer will either increase or decrease the range, depending on its instantaneous setting.

Hold Off allows another function permitting better displays of aperiodic signals such as complex digital or tv analog waveforms. In this way, the time between two alternate sweeps may be controlled so that triggering during this interval is impossible. Once again, however, brightness may be reduced, so an optimum conditon should be selected. There is also a single sweep function, permitting a 1-time event to trigger the sweep generator and be photographed.

An X-Y amplifier function is also included whereby the two vertical Y channels with identical bandwidths and impedances may process incoming waveforms at quadrature (90°) and display them in-phase at right angles, or out-of-phase at quadrature, or conditons in between. This is how very useful vectors are produced to evaluate phase angles and the highly important chroma circuits of color television receivers. To do this, a very clean gated rainbow generator is used rather than one generating standard NTSC excitations. NTSC evaluations depend on levels which can be easily shifted.

External sync inputs are available, too, for those impossible signals that refuse to sync on standard internal scope operations. A jittery waveform such as one originating from automobile ignition high voltage might be a good example of this. For a steady display you'll have to use external sync from the engines No. 1 cylinder. And do be careful of its amplitude, using only a loose coupling about the spark plug wire lead as your source. Otherwise, the 30 or 40 kilovolts will blow the scope all the way back to its factory.

External sync is only necessary when viewing low frequency waveforms having considerable movement caused by unstable generation. Usually, kilohertz frequencies and above almost never require external sync, because internal sync has improved markedly in all oscilloscopes over the past five or more years, along with increased bandwidths and calibration accuracies. When using external inputs, always pay attention to the scope's input impedances and waveform input amplitudes. You don't want to load the instrument nor short your wave source. Both are possible with mismatched impedances and dubious current sources.

Instrument instruction books are meant to be read and heeded. Those who don't do their homework put both their instruments in jeopardy as well as their testing abilities. I suggest intelligent attention to such details.

Vertical Amplifier Delay Lines are what you will not discover in lower-priced oscilloscopes. Such added attractions cost money and unless you intend to view everchanging waveforms, rather than recurrent ones, you won't need this slight extravagance. However, should you be required to either trigger on marginal signals or evaluate odd-ball voltages, scopes without vertical delay lines are usually quite acceptable, especially if they have the necessary low and high pass filters coupled with holdoff.

If there is any doubt, try before you buy. Then you'll be sure your needs are satisfactorily accommodated. Cheap instruments that don't do the job are the world's worst waste of money. But expensive instruments, especially with counters and digital voltmeters integrated into a massive box, are unduly expensive and difficult to service without accumulating marvelous results.

Always know your instrument's advantages and limitations, then you'll escape the inevitable unpleasant surprise. Multifunction packages have their places, but probable downtimes are a powerful consideration since most readings are taken through a common transducer and affect the entire equipment's operation.

Spectrum Analyzers

Here is another breed of electronic analyzer that's totally different in operation from the conventional oscilloscope. But with higher frequencies being made available and cochannel and adjacent channel interference and radiation of such concern, these spectrum analyzers are invaluable. Pricewise they're expensive, and $15,000 to $100,000 cannot be considered unusual, depending on frequency range, digital storage, programmability, and sensitivity. Tektronix and Hewlett-Packard are two of the industry's leaders in the manufacture of quality spectrum instrumentation. Several Tektronix modules are shown in Fig. 6-5.

A spectrum analyzer is probably best described as a swept front-end superheterodyne radio with detector. It will measure waveform amplitudes in dB and linear, and frequency in terms of Hertz to Gigahertz, depending on design and control settings. Resolution of any voltage measured extends from Hz to MHz, with each analyzer calibrated selectively by the manufacturer.

Sensitivity is often as much as -115 dB down, with resolutions of as little as 30 Hz not uncommon. Markers of one description or another are usually available to identify various frequencies and their desirable or undesirable states in question, and sweep time divisions may appear in the calibrate position when used for frequency domain displays and between 5 msec and 1 μsec for variable

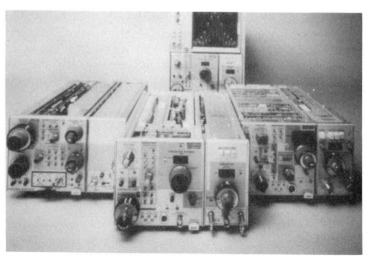

Fig. 6-5. Typical Tektronix spectrum analyzer plug-ins covering dc to 18 gHz. (courtesy Tektronix)

rates to stabilize the display for maximum ease of viewing. These time bases also have an amplifier position that's used with external sweeps when required.

There is also a Frequency Span per division selector that's normally coupled with a Resolution selector so that the two will remain within the instrument's calibrated ranges. But there is a Variable Frequency Span arrangement that uncouples this selector so it may be used independently, although it's rather easy to exceed instrument ranges. When this occurs, a symbol appears on the graticule readout indicating an out-of-calibrate condition, usually in the form of a two-sided triangle with the bottom left open; or, in mathematics language, a greater-than sign signifying overrange.

The one very tender part of any spectrum analyzer is its mixer input or front end. It can accommodate very little dc voltage and, above a few millivolts, under ordinary circumstances, you can damage the initial electronics severely, resulting in a repair bill that can easily amount to hundreds and possibly thousands of dollars, depending on the quantity of overload.

Consequently, always try and use either a blocking capacitor or a 75/50-ohm conversion impedance matcher that has a 5.72 dB low loss characteristic in addition to the capacitor if an impedance match becomes necessary. As most of you are aware, microwaves and digital logic are normally designed for 50 ohms while CATV and video installations use 75 ohms. Should these impedances remain mismatched, standing waves (VSWR), ringing, power loss, and all sorts of unpleasant results can become apparent as frequencies increase and decrease. At low frequencies such as kilohertz and below, impedance matching isn't overwhelmingly important, but at megahertz and gigahertz, precision matching becomes an absolute must. Even 3 and 6 dB padders are used at GHz to avoid even an appearance of such unpleasant affects that could easily distort measurements within any accurate dB range.

Improper cabling is another perennial problem associated with high frequencies which often requires waveguide assemblies and other carefully engineered fittings. And as we move more and more into the high megahertz and gigahertz spectrums, both proven and advanced techniques must be used to process signals about the earth and in outer space. All this, of course, results from constantly increasing demand for expanded communications services that soon will span both the U.S. and the world. Be it video, audio, or data, there's no end in sight for additional spectrum requests since so much in communications amounts to big bucks, indeed!

A Modern Analyzer

Let's take the front end of the modern Tektronix 7L5 spectrum analyzer and work through the block diagram in Fig. 6-6 to give you an idea of how it works. And once we're through with this description, then you will have some appropriate applications to view that should furnish at least a general idea of what is done with this type of analyzer in 2-way radio.

With inputs to the 7L5 of 0 to 5 MHz, the main oscillator is both tuned and swept between 10.7 and 15.7 MHz, producing an intermediate frequency (i-f) of 10.7 MHz. Meanwhile, the frequency of this main oscillator derives from A and B oscillators in the 11 to 16 MHz range feeding into a phase lock comparator that locks their frequencies. A sweep frequency control sets these oscillators according to manual front panel settings of the Dot Frequency and Frequency Span controls. In turn, a 10 MHz master oscillator having suitable divide-by counters that produces output references of 500 kHz, 100 kHz, and 10 kHz, for A and B oscillators as well as the 2nd Local Oscillator that operates at 10.45 MHz.

The Second Mixer then produces a heterodyned output of 250 kHz, and this i-f is then processed through a variable resolution filter for bandwidths of from 10 Hz to 30 kHz. Afterwards, the signal is once more amplified, then detected, and video proceeds through additional amplifiers for the 10 dB, 2 dB, and linear outputs. Further video proceeds to display processors where it is either stored or produced on the face of the cathode ray tube. When Display A or B latches are engaged, the information converts to digital and is stored in either A or B memory, then converted back to analog and routed through additional amplifiers to the mainframe. Vertical (video) signals may be averaged or peak detected for CRT display at selected rates. Phase comparators and loop filters produce error voltages that maintain all the necessary frequencies via the phase lock loop comparator. The Frequency Span/Division determines the A/B oscillator sweep and horizontal voltage via an attenuator and binary switch to a summing point. A ROM looks at any selected span and chooses one of three sweeps outputs from the binary switch.

We could, of course, go into considerably more detail but this portion is simply a primer introduction so that the following applications won't appear out of the blue as a rude shock. Some know nothing of spectrum analyzers and this is just our way of showing how and why these complex and very expensive instruments operate. In effect, a spectrum analyzer electronically converts some

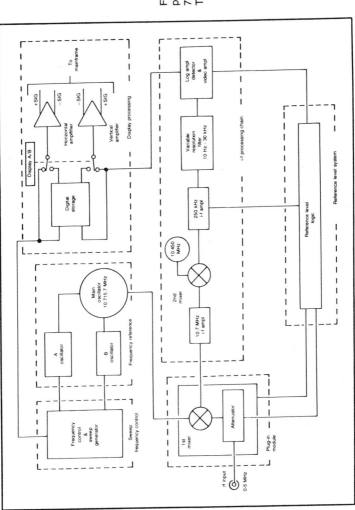

Fig. 6-6. 7L5 spectrum analyzer plug-in (dc to 5 MHz) for Tektronix 700 series main frames. (courtesy Tektronix)

Fourier transform operation into a viewable signal having both amplitude and frequency components as you will shortly see. And it is one of the most practical and usable measurement techniques in both video and 2-way radio. For instance, can you offer another simple way to measure modulation, sidebands, spurs and harmonics, or identify interfering frequencies and cross modulation?

A broadcast signal. Just to get a feel for a spectrum analyzer's ordinary mode of operation, let's take a video signal directly off the air, using the 50/75-ohm impedance matcher with a 5.72 dB loss. The signal is shown in Fig. 6-7. This, of course, is added (+) to the negative dBs shown for a true readout. But when we move on to an actual transceiver, then 50 ohms with no intervening RC network applies all the way. And since the TV broadcast band begins at 54 MHz, we'll use a 7L12 analyzer plug in our 7613 mainframe.

To begin the series, we'll dial a center spectrum reference of 184 MHz (±10 MHz) and look on either side for the obvious signals appearing there. As the scope readout shows, each division is worth 2 MHz in this measurement at a resolution of 300 kHz for each. Since the video carrier is always 4.6 MHz lower in frequency than the audio carrier and is amplitude modulated, the initial carrier appears at 8 MHz less 184, or 176 MHz, and that's very close to the

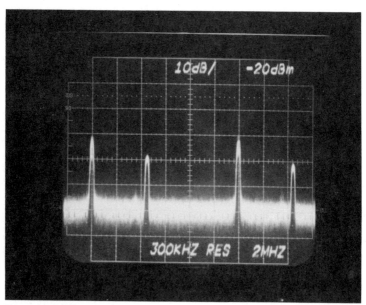

Fig. 6-7. Looking at the video and audio carriers of two tv stations. Analyzer sweep is centered at 184 MHz.

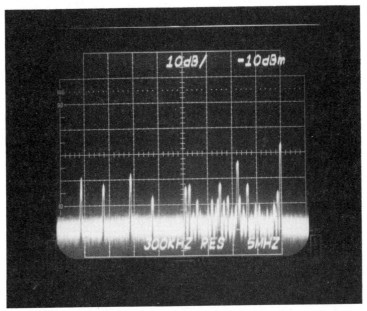

Fig. 6-8. Maritime, amateur and land mobile show up well on spectrum displays.

given 175.25 MHz for Channel 7. The next video carrier resides at 4 MHz above 184 MHz, or 188 MHz, and that's quite close to the 187.25 MHz assigned to Channel 9. So, immediately, we've identified those two frequencies quite easily and see that they are operating as scheduled with a signal strength of −20 dBm +10 dB/div. +5.72 dB. The two video carriers are at approximate transmit/receive levels of −52 dB +5.72 dB, or −46.28 dBm; provided, of course, your analyzer has been calibrated correctly before beginning these measurements. Note that relative levels are in dB, while the final, precise level is stated in dBm, just as indicated initially on the graticule. These, then, are decibels referenced to one milliwatt.

In Fig. 6-8 these frequencies are taken down a notch or two into the frequency modulation domain. Here, the center frequency has been reduced to 90 MHz, and each vertical division amounts to 5 MHz. Recalling that FM stations are 200 kHz apart, and television stations are separated by 6 MHz, you immediately note that just about 20 MHz below center frequency you will find video carriers for Channels 4 and 5, while above 90 MHz is the FM band that extends to 108 MHz. So both low band tv and the entire FM spectrum are included in this one display, whose highest amplitude signal is −46 dB + 5.72 dB, or −40.28 dBm. Of course some of the radio signals are as far down as −80 dBm. So you may need a fairly sensitive tuner to track them all.

The waveform in Fig. 6-9 has its 0 center frequency moved over to the left and the "blips" that show are identified at 800 kHz, 1.1 MHz, 1.2 MHz, and 1.5 MHz - all within the AM radio band, and all exuding amplitude modulation like crazy. Note these are strong signals, beginning at +20 dB, but three of the four have relatively poor signal/noise ratios and could be problems for good reception. Note, too, that you're having to take our word for the 30 kHz resolution and 200 kHz/division readouts. Here, a very low value bandwidth filter would have killed the display, so we had to include all the noise, like it or not, blanking out regular spectrum parameters. Observe also that we're talking about carrier-to-noise not signal-to-noise which are entirely different parameters, even though we don't have to include the usual bandwidth factor which, for this analyzer, amounts to about 250 kHz when using the video filter. But if this filter was used, the correction factor would amount to 12 dB for a 4 MHz bandwidth. Subtracted from 45 dB, which is the apparent measured C/N, the real C/N would become 33 dB, and

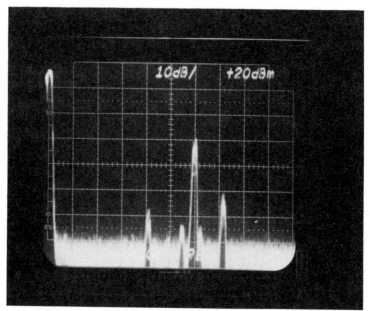

Fig. 6-9. Easily identifying frequencies at 0.8, 1.1, 1.2, 1.5 MHz when displayed at 200 kHz/division.

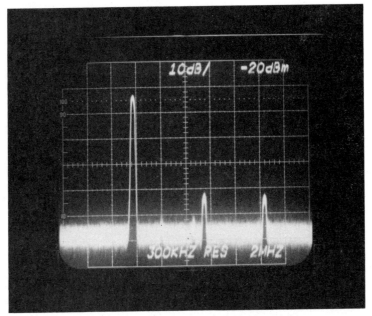

Fig. 6-10. A 27 MHz carrier with its second and third harmonics.

that's excellent for the more prominent of the several signals ... The others being not nearly as well off. Obviously this is a local station with a whopping output since the top of the graticule is referenced at +20 dBm. There should be no trouble hearing this one all over the countryside.

Now that you have an introduction to the general usage of a spectrum analyzer off the air in tv, FM, and AM applications with their appropriate modulation, you should be ready to transfer over to some of the measurements that are critical but relatively easy to make in 2-way radio. I'll demonstrate both AM sideband as well as single sideband transmissions just to keep you up to date.

At about 10 meters and also around 15 meters, maritime, amateur, and land mobile operations abound. So I thought a 27-30 MHz carrier and its second and third harmonics might be of interest, especially if a couple of spurs were evident in between. Figure 6-10 is nothing more than keying a transmitter into an attenuator block, and thence into the spectrum analyzer, taking care that the coupling is well-supplied with a dc blocking capacitor and that sufficient attenuation exists to prevent any ac overload. Next, in Fig. 6-11, I place the reference on the left graticule vertical line and adjust for 10 MHz/division. The carrier appears at 27 MHz, followed by a couple of interesting developments. These are spurious voltages measured at some 37 dB down below the carrier, and within 6 MHz and 12 MHz, respectively, of the carrier. The output of this transmitter, of course, probably can be tweaked to put these at −40 dB below the carrier where they should be. The second harmonic now shows on the right at some 54 MHz, and is 23 dB down from the carrier. The third harmonic, on the far right, appears at 82 MHz, and has a relative amplitude to the carrier of about the same as the original pair of spurs.

If you'd care to have this spur-to-carrier relationship expanded, analyzer controls are reduced to 2 MHz/division and you not only see the two spurious voltages plainly, but there are another two spurs at 2 and 5 MHz between said spurs and the carrier, although much lower in amplitude than the originals. Note that the original −20 dBm sensitivity and 300 kHz resolution have been maintained for these examinations and only the frequency/division factor has been changed. If you wanted carrier-to-noise and used the 30 kHz filter, allowing for a 3 kHz passband, there is about a 0.3 dB subtractive factor and you'd have about a 50 dB C/N reading, which seems almost too good to be true. (C/N = 10 log signal BW/filter BW + 2.5 dB).

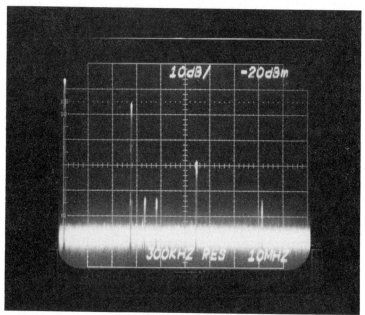

Fig. 6-11. The carrier, spurs, and other intruders.

At any rate, if you're curious about single sideband—upper single sideband, that is—Fig. 6-12 will show what to expect. This is now peak envelope detection, however, rather than just a carrier output measurement, so the signals are obviously different and appear in very separate segments compared with standard AM. Once again, we used the 30 kHz filter if you're interested in making further examinations of the signal in C/N.

But what about power readouts from these measurements. These can be made, also, with a spectrum analyzer, and are particularly important in both microwaves and satellite signal analysis. Once again, you're working with logarithms because of the very large or very small numbers involved. The usual method for negative numbers, however, doesn't hold since you'll have to translate these into positive units by first dividing them by the 10 of 10 log, and then subtracting from 10. This immediately adds a further power of 10 to the final answer, and so care is needed in handling such a power equation.

As you should know, $\text{Power}_{dBm} = 10 \log P_2/P_1$, where your reference of P_1 equals 10^{-3} watts. Therefore if you are using negative dBm values such as -20 dBm, you'll have to do a little equation rearranging:

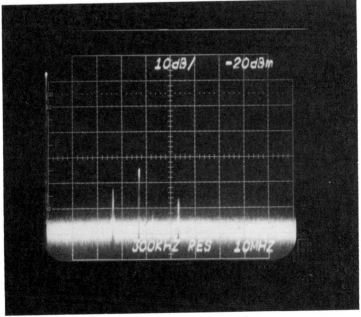

Fig. 6-12. A single sideband display.

$$\text{Log } \frac{P_2}{P_1} = \text{antilog } \frac{(\text{Power in dBm})}{10} \text{ and}$$

$$P_2 = \text{antilog } \frac{(\text{Power in dBm})}{10} \times P_1$$

As a little demonstration, let's evaluate the allegedly suppressed single sideband photo with its rather poorly suppressed carrier. Here, the sideband is −64 dB which is down in the mud, indeed.

$$\frac{64}{10} = 6.4 - 10 = 3.6 \times 10^{-10}$$

Antilog of $3.6 = 3.98 \times 10^{-7} \times 10^{-3}$, or 398 picowatts (pW)

Then multiply by 1.414 for peak power.

As you might expect, this signal was extracted through a great deal of attenuation, and is suitable in this range for both microwave and satellite-level considerations. In fact, we expect to be using a spectrum analyzer exclusively for both frequency and power measurements in the 4 GHz spectrum. You might also like to know there are dBm to dBW conversion tables available where math is not required for less than precise calculations. On the other hand, a good range with relative accuracy for spectrum analyzers does, indeed, exist through the 4 GHz downlink frequency, so you may want your microwatts and milliwatts to be as precise as possible. There are also calculators offered with negative signs that will read out the required antilog without having to use the awkward mantissa. Simply divide your 10 into the negative log, as before, insert the minus sign, then take the antilog, and the number comes out exactly.

If this is slightly confusing, let's do another example just as it would appear on your calculator, but this time using the AM illustration, remembering that we're measuring carrier power only in this instance.

The dBm reading, therefore is a −35, since our reference is −20 dBm to begin with. The equation now becomes the following.

$$P_2 = \text{antilog}\left[\frac{\text{Power (dBm)}}{10}\right] \times P_1 \text{ (in milliwatts)}$$
$$P_2 = \text{antilog } -3.5 \times 10^{-3} = 3.1622 \times 10^{-4} \times 10^{-3}$$
$$P_2 = 316 \text{ nanowatts}$$

And in case you're really interested, our attenuation from input to output of this padder amounted to some 70 dB, which translates into 10^7. So $10^7 \times 316 \times 10^{-9}$ equals 3.16 watts, and that's what this particular transceiver was emitting into its 50 ohm loaded output. As many well know, there are lots of ways to whistle Dixie.

One further reminder; as frequencies increase in the gigahertz (10^9), spectrum analyzers usually decrease in accuracy and power meters are often preferable when extraordinary measurements are required. However, as further experience with analyzers in extended spectrums continue, we would expect some improvements fairly quickly. Therefore, before you buy, know intended and extended measurement parameters and purposes since spectrum analyzers usually appear only under the more affluent Christmas trees.

The foregoing should be about all the tips you'll need in oscilloscope usage around 2-way radio sets. Naturally, things like modulation and carrier displays are interesting, but these are straightforward and those who will be working on cellular should have this background already wrapped up. One final illustration we'd like to show consists of a bunch of AM signals jammed together in the lower frequency end of the spectrum like those in Fig. 6-13.

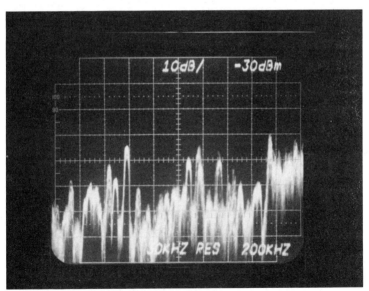

Fig. 6-13. Taking AM signals off the air at 30 kHz resolution and a lot of fast-moving modulation.

By spreading display ranges and increasing resolution, interferences are easily spotted and trapped, wherever necessary. This example is just another reason for additional, intelligent use of the spectrum analyzer. Naturally, this applies to satellite and microwave signals, too. With an appropriate receiving antenna—a 4-foot dish, for instance—microwave interference can easily be identified and, hopefully, dealt with.

MOTOROLA DIAGNOSTICS

The base site controller (BSC) continuously monitors base station alarms and equipment performance by either a terminal at the base station location or remotely through the EMX. Faults reported to the EMX automatically reconfigures the base station so that transmissions and records remain on the air and are properly stored. As you may well imagine, there are numerous diagnostic tests, both automated and manual that are available, and I'll try and touch base with most of them, but not include every excruciating detail. Other manufacturers besides Motorola, of course, will have variations of these schemes, but the general approach could be fairly universal and a reasonable description of this system should offer a fair understanding of them all. Unfortunately, at this time, a number of would-be makers of cellular equipment either haven't progressed far enough in their design or are simply overly reluctant to talk. I suspect the former is more fact than fiction.

Initially, in nonredundant systems, the defective subsystem is removed from service, reducing signal transmission and/or reception. Where there is redundancy, a backup takes over and restores communications to their normal states. Later, a substitute module replaces the faulty one upon technician exchange. There are also manual system tests initiated from the EMX for the base system controller. These include:

☐ Status routines which check channel controls, handoffs, sector selections, and scan reports.

☐ Controls such as redundancy and switchover, insertion or deletion of voice channels, signal channel on or off, and scan receiver frequency and antenna selection.

☐ Listing in-service and failed modules.
☐ Audio loopback.
☐ Mobile test routines.
☐ Setup, BSC, and rf commands.
☐ BSC and rf exercises.

Site Control Processor

This unit deals with automated reaction to faults in the system by doing the following: it retains a record of alarms; returns the base site to normal after a failure correction; monitors keyed and command transmitter states plus power output; triggers high temperature alarms for rf power amplifiers; automatically deletes voice channels with 2nd injection lock alarm, exciter or low power alarms, automatically places channels in or out of service with insertion or removal of audio conditioning boards; initializes processor boards when inserted; initializes voice channels via EMX commands; with BSC, maintains statistics on SAT, 10 kHz signaling tone, and short calls; and inserts any redundant modem data link, as well as signal channel processor, or scan processor via BSC reconfiguration.

The Processor also tests itself for, modules affected, ROMs and RAMs, digital loopback, A/D converter, 3 MHz reference, and

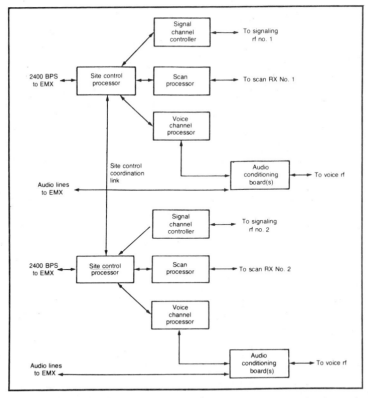

Fig. 6-14. Motorola's redundant 16-channel BSC block diagram showing multiinputs and outputs as well as signal flow. (courtesy Motorola Communications)

FIFO. A redundant 16-channel BSC configuration is illustrated in Fig. 6-14. Here the BSC connects to the EMX through a 2400 bps modem, with an identical modem line added to offset line failure. Since only 1 line is used at a time, the BSC will switch lines automatically when there is a fault. The site control processor (SCP) controls one signaling channel controller, one scan processor, and up to two voice channel processors. And this arrangement permits a total of one scanning receiver, one signaling channel, and 16 voice channels. This may be expanded by duplicate equipment additions.

Rf Alarms include the 3 MHz oscillator, 15 kHz oscillator, SATs, transmitter, receiver, injection locks, high power, power level, ac failure, and other spare alarms.

Base Site Controller Alarms involve software queue full, scan processor response time, excessive SAT failures, processor modules inserted or removed, voice channel processor failure, SAT distribution failure, and 3 MHz distribution failure.

Statistical indicators involve antenna usage, inward trunk problems, outgoing trunk problems, early call terminations, and excessive or increasing SAT failures.

In redundant BSC operation voice channels are divided between two BSCs, and no single fault will remove the site from service, but there can be some degradation. Motorola, therefore, recommends two data lines be installed so that no single failure stops service. In nonredundant configurations, the site control processor and/or data line and power supply can put the site out of service. The same is true for the signaling channel controller—the calling channel would vanish from the air, although mobiles might still originate through an adjacent cell. Scan processor failure would affect handoff ability, mobile status updates, and antenna reassignments. If the voice channel processor (VCP) goes out, all voice channels so controlled would be lost.

Improved Error Detection with Test Mobile

Signaling channels, scan receivers, and the various voice channels may be tested by scheduled calls using this method. And undoubtedly, this method will be extensively used in new installations and those without redundant automation. Well established cell sites undoubtedly will rely largely on EMX probes and automated test procedures to check early failures. But it will always be the human technician who has to make modular replacements and observe final test results. He'll be hard to replace, redundancy or not!

Fig. 6-15. ITT-Pomona's excellent in-line IC DIP removers and test clips.

ELECTRONIC ACCESSORIES

Sometimes it's the little things that count, even in the world of superfrequency electronics—things such as microwave connectors, patch chords, adapters, and especially integrated circuit test clips. Whether analog or digital, signals may always be observed at the various pinouts of any IC under operational conditions if that IC can be monitored without a probe slip shorting those tender silicon chip electronics. In my investigations, I came across ITT Pomona's series of dual in-line DIP removers and test clips. The latter have either tarnish-resistant nickel-silver contacts or gold-plated beryllium copper contacts for considerably longer life and lower contact resistance than many of their competitors. In addition the clip molding material is remarkably sturdy and heavily sprung so that good contact is readily and securely established. Available in 8- through 40-pin versions you'll find them easy to handle, reliable, and very useful in those final troubleshooting stages and assessments when repairing either radios or test gear. Figure 6-15 shows the 40-pin model. We commend them to your use.

Similarly, we would also heartily recommend Powermaster's Line Monitor Power Conditioner, shown in Fig. 6-16. Test bench

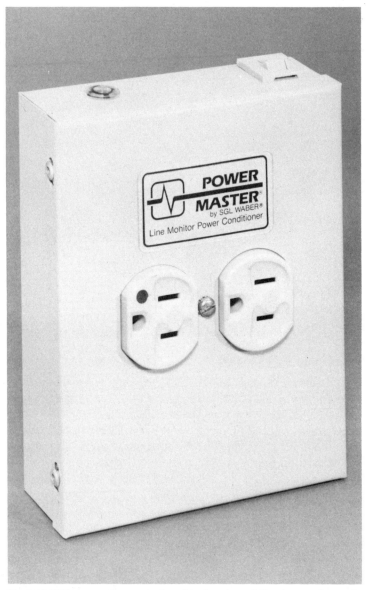

Fig. 6-16. With new and expensive bench test equipment, it makes good sense to protect both radios and instruments with effective spike and noise suppression by SGL Waber and Powermaster.

and equipment power leads are often loaded with high voltage spikes, overvoltages and highly undesirable noise—the very enemies that can permanently damage a $40 thousand spectrum

Fig. 6-17. Bird's new 4410 wattmeter, offering 18 rf connectors and measuring power levels of 0.25 W to 10 kW at frequencies between 200 kHz and 1 GHz.

analyzer, a $12 thousand test set, or a $5 thousand base radio. In factories, computer terminals, office equipment locations, and other highly active areas, these conditions positively do exist. And for $100 or so you can do so much to prevent ac line catastrophic effects. As Powermaster correctly states, with its line monitors and conditioners you may not need a dedicated line, a voltage regulator, or extremely expensive ac line filters designed especially for your equipment. SGL Waber, the manufacturer, has more than 150 stocking locations that carry a complete line of power strips and spike suppression equipment. My own test benches are thoroughly protected with two LM 3100 plug-in units that carry up to 15 amperes each, offer overload circuit breaker protection, a 7-stage filter network for voltage spike and noise reduction or elimination, and a pair of male load plugs for easy wall-socket insertion. Two outlets for a spare and the workbench do the rest. You'll like a 7 kilovolt spike reduced to 55 tiny volts, noise rejection attenuation of 20 to 40 dB, and a clamping response time of less than or equal to 10 nanoseconds. It does make sense, and pretty inexpensive insurance at that, especially when you're on the move these mobile filters can move with you.

Of course, if you intend to make separate, possibly more accurate power measurements, don't forget your friendly meter manufacturer, Bird Electronics. They're so closely identified with 2-way radio that the name's probably wholly familiar, even without explanation. However, Bird's Herb Heller does want you to know of a new and relatively inexpensive THRULINE® directional wattmeter for CW and FM systems that handles (with plug-ins) 0.25 W to 10 kW in systems with frequency ranges between 200 kHz to 1 GHz. Fully portable, ranges are selected by a front panel rotary switch, and elements are rotated for either forward or reflected power measurements. Insertion VSWR is only 1.05 (or less). There are 18 common rf connectors available, and overrange protection is 120 percent. The Model 4410 in Fig. 6-17 sells for $495 and the plug-ins range from $125 to $175.

Index

A

Access, 15
Access channel, 93
Alarm center, 45
Alarms, 44, 207, 208
AMPS, 2, 7, 14, 16-19
AMPS in Chicago, 28, 35
Anaconda-Ericsson, 123
Analog color code, 93
Answering, 15
Antenna, transmit, 26
Antenna gain, 13
Antennas, 23
Antennas, location of, 9, 10
Antennas, MSAT, 56
ARTIS, 160, 161

B

Base station, Motorola, 106
Base stations, 32, 129
Baud rate, 134
Baud rate, mobile, 16
Baud rate, MTSO, 17
BCH code, 93
Billing, 7, 19
Blank and burst, 20, 21
Busy-idle bits, 93

C

Calling, 68, 136
Calling, PRCS, 157, 158
Calling rules, PRCS, 163
Call processing, mobile, 68-75
Carey contours, 36
Carey Curves, 41, 42
Cell makeup, 20
Cell radius, 39
Cell site controller, 134
Cell site duties, 19, 20
Cell site overlap, 35
Cell sites, 8-14
Cell splitting, 13, 40
CELLTREX, 146-148
Cellular junior, 153
Cellular satellite, 45-60
Central processing unit, 138
Central processor, 101
CGSA, 37
Channel, audio, 22
Channel, setup, 22
Channel assignments, 31
Channel confirmation, 71
Channel frequencies, 23
Channels, mobile, 65
Channels, number of, 8, 52
Channels, talking, 22
Channels, voice, 22
Channels for PRCS, number of, 156
Charges, 7
Class, mobile, 64
CMS 8800, 123
Compatibility, 3
Control channel, 93
Control message, mobile, 82

213

Control unit, OKI, 121
Conversation, mobile, 74

D

Data restrictions, 90
Default, PRSC, 165
Definitions, 93
Diagnostic aids, 44
Diagnostics, 141
Diagnostics, mobile, 208
Diagnostics, Motorola, 206-208
Diagnostics, remote, 185
Digital color code, 93
Disaster communications, 59
Disconnect, 16, 18
Disconnect, PRCS, 165
Disconnects, 137
DMS, 132, 138
DYNA TAC, 99-114
DYNA TAC base station, 106
DYNA TAC mobiles, 110
DYNA TAC portables, 110

E

E.F. Johnson, 148
Electronic mobile exchange, 100
Emergency medical communications, 59
Emissions, limits on, 62
Equipment approval, 35
Error detection, 208

F

Fading, 23
Falsing, 21
Faults, system, 44, 207
FCC, 23, 27, 45, 59, 64, 155
Features, 138, 143
Features, mobile, 6
Firmwave, 106
Flash request, 94
Forest fire reports, 59
Format, forward control channel, 81
Format, forward voice channel, 92
Format, overhead message, 84
Format, PRCS message, 163
Format, PRCS signaling, 170
Format, reverse control channel, 75
Format, reverse voice channel, 79
Forward control channel, 81, 94
Forward voice channel, 92, 94
Frames, 20
Frequencies, reusable, 29, 38, 41, 43
Frequencies, satellite, 47, 50, 52
Frequency allocations, 31, 33
Frequency pairs, 65
Fujitsu Ten, 149

G

General Electric, 130
General Electric-Northern Telecom System Features, 138
Group identification, 94
Group multiplexer unit, 102

H

Handoff, 10, 18, 94, 134, 137
Handoffs, 2, 14, 24, 137
Hazardous materials transfer, 59
Home mobiles, 14
Home mobile station, 94
Housekeeping, 19

I

Identification, mobile, 66
Identification, PRCS, 160
ID number, mobile, 17
I/O hardware, 102
ITT's CELLTREX, 146-148

L

Land calls, 33
Land party calling, 136
Land station, 94
Lay enforcement, 59
LMSS, 45, 50, 57
Locating, 14
Locations, base station, 32

M

Maintenance and Status unit, 102
MAP diagnostics, 141
Microcomputers, 101
Microprocessor, 101, 207
Microprocessor control, 20
Microprocessors, 10
MIN 1, 66
MIN 2, 66
Mobile awaiting answer, 73
Mobile calls, 33
Mobile features, 143
Mobile identification, 66
Mobile identification number, 94
Mobile-mobile calls, 33
Mobile-mobile party calling, 136
Mobile operations, 17
Mobile party calling, 136
Mobile radio, trunked, 146
Mobile radios, 142
Mobiles, 111
Mobile station, 94
Mobile telephone exchange, 37
Modems, 134
Modularity, 20, 132

Modulation products, 63
Motorola, 99-114
MSAT, 47, 53-58
MTSO, 8, 14, 16, 18, 19, 33
MTX, 132, 137

N

NASA, 45-60
Northern Telecom, 130
Numeric information, 94

O

OKI Advanced Communications, 114-123
OKI receiver, 119
OKI transmitter, 115
Operator duties, 44
Orders, 97
Oscilloscopes, 186
Overhead information, 70, 71
Overhead message, 15, 84

P

Paging, 15, 17, 97, 134
Paging channel, 97
Parameters, PRCS, 168
Personal Radio Communications Service, 153
Portables, 114
Power loss, 20
PRCS, 153
PRCS parameters, 168
Price, PRCS, 153
Prices, 5, 52
Propagation losses, 23
PROTEL software, 133, 140
Protocol, PRCS, 162

Q

Quality of transmission, 6

R

Radio Common Carrier, 2
Radios, location, 22
Radios, setup, 22
Receiver, land/base, 64
Receiver, OKI, 119
Receivers, mobile, 64-68
Receiving, 15
Receiving, mobile, 69
Receiving rules, PRCS, 164
Redundancy, 17, 101, 206, 208
Registration, 97
Release, mobile, 75
Release request, 97
Repeaters, 156, 162, 166-168

Repeaters, licensing of, 159
Reverse control channel, 75, 97
Reverse voice channel, 79, 97
Rf signals, 64
Roamer, 97
Roamers, 14
Roaming, 128

S

SAT, 18, 19, 67
Satellite links, 57
Satellite system, mobile, 49
Scan of channels, 97
Search and rescue communications, 59
Seizure precursor, 97
Service request, mobile, 72
Shared memory shadowing, 101
Signaling, PRCS, 161
Signaling formats, 75-98
Signaling tone, 98
Signal-to-interference ratio, 13
SMSA, 4, 14, 58
Software, 105, 133, 140
Specifications, cellular, 61-98
Specifications, PRCS, 169
Specifications, transmitter, 64
Spectrum analyzer, Tektronix 7L5, 196
Spectrum analyzers, 194
Standards, PRCS, 160
Stationary radios, 142
Station class mark, 67
Statistical indicators, 208
Status information, 98
Subscribers, maximum number of, 6
Subscribers, number of, 16, 52
Supervisory audio tone, 16, 98
Supervisory signals, 14, 16
Survey, opinion research, 154
Switch unit, 102
System concepts, 7
System identification, 98
System operation, PRCS, 156

T

Talking channels, 22
Test equipment, 185-212
Test procedures, automated, 208
Test programs, 185
Time bases, 188
Tone signaling unit, 102
Traffic analysis, 43, 208
Transceiver, 22
Transmission lines, 26
Transmitter, land/base, 63, 64

Transmitter, OKI, 115
Transmitter, WECO, 30
Transmitters, 40F9, 30
Transmitters, 40F9Y, 30
Transmitters, mobile, 64-68
Trucking, 58

U

User market, 58
User market, PRCS, 154

V

Voice channel, 98
Voice signals, 63, 64

W

Western Union in Buffalo, 35-45
Wideband data, 63
Wireline system, 27, 28